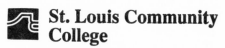

MODERN AUDIO TECHNOLOGY
A Handbook for Technicians and Engineers

Martin Clifford

PRENTICE HALL, Englewood Cliffs, New Jersey 07632

Library of Congress Cataloging-in-Publication Data

Clifford, Martin, 1910-
    Modern audio technology : a handbook for technicians and engineers
/ Martin Clifford.
        p.   cm.
    Includes index.
    ISBN 0-13-051889-1
    1. Sound--Recording and reproducing.   I. Title.
TK7881.4.C55   1992
621.389'3--dc20   91-44574
                                                                         CIP

Editorial/production supervision
  and interior design: Carol Atkins
Cover design: Lundegren Graphics, LTD.
Manufacturing buyer: Susan Brunke
Prepress buyer: Mary McCartney
Copy Editor: Linda Thompson
Acquisitions Editor: George Kuredjian

© 1992 by Prentice-Hall, Inc.
A Simon & Schuster Company
Englewood Cliffs, New Jersey 07632

The publisher offers discounts on this book when ordered
in bulk quantities. For more information, write:

Special Sales/College Marketing
College Technical and Reference Division
Prentice Hall
Englewood Cliffs, New Jersey 07632

Printed in the United States of America

10 9 8 7 6 5 4 3 2 1

ISBN   0-13-051889-1

PRENTICE-HALL INTERNATIONAL (UK) LIMITED, *London*
PRENTICE-HALL OF AUSTRALIA PTY. LIMITED, *Sydney*
PRENTICE-HALL CANADA INC., *Toronto*
PRENTICE-HALL HISPANOAMERICANA, S.A., *Mexico*
PRENTICE-HALL OF INDIA PRIVATE LIMITED, *New Delhi*
PRENTICE-HALL OF JAPAN, INC., *Tokyo*
SIMON & SCHUSTER ASIA PTE. LTD., *Singapore*
EDITORA PRENTICE-HALL DO BRASIL, LTDA., *Rio de Janeiro*

Dedicated to
Harry and Betsy Chapman,
in Friendship

Contents

Contents

CHAPTER 9   SURROUND SOUND   237

Preface

Audio has changed from a passive art or science to an active topic of interest for a diversity of music lovers, ranging from beginning musical hobbyists to sophisticated studio professionals. In effect, the "new music" has something for everybody. At one time those who loved music became infatuated with equipment specs, looking for electronic components that could bring the concert hall into the home. Today they can indulge their creativity, for they can reproduce not only the sound of a single instrument but that of an entire group, with the capability of creating completely new sounds and completely new rhythms. And that is the reason for this book—to explain what can be done with sound, how to generate it, how to modify it, and how to record it.

Sound is so closely interwoven with living that learning about it might seem unnecessary. The purpose of Chapter 1, "Elements of Sound," is to dispel that illusion. Sound is the stuff of which audio is made. It is essential to have an understanding of the raw materials.

Chapter 2, "Electrical Noise," might seem to be wholly unnecessary at first thought. While some noise is undesirable, some, especially noise that has a periodic quality, can and does contribute to the enjoyment of music. There are instruments available whose sole purpose is the deliberate introduction of noise as part of a music score. Noise can be pleasant when it is subject to control.

To be able to produce sound is not enough in itself. It must be accompanied by the ability to communicate data to others so that they can also generate the same sound and do so precisely. Sound is much more than a vague rendering of tones in a certain order. Sound must be measured; to be able to do so on a repetitive basis, a reference, a starting point, must be supplied. There are a number of available references; for sound measurements to make any sense, these must always be specified. Not only are there various references but measurement units as well. The purpose of Chapter 3 is to remove the mystery and confusion surrounding them.

Musical recording and playback starts with a microphone, a transducer-type device that changes sound energy to electrical energy. This conversion permits the manipulation of sound, now in the form of an electrical voltage, so that it can be changed, increasing it or decreasing or altering its shape if necessary. Ultimately, though, as explained in Chapter 4, that electrical voltage must be changed back into sound.

During the time that sound exists in voltage form, it is manipulated, limited, expanded, trimmed, and modified to suit specific needs. This can be done by relatively simple devices, such as filters, pads, and attenuators. As explained in Chapter 5, a filter is an electronic traffic control, permitting some electrical signals to continue on but blocking others. The electrical sound signal is amplified by solid-state devices, with the increased signal strength carefully controlled by components such as pads and attenuators.

At one time all audio signals were analog, with signal recording and processing in that form. Substantial control over audio became possible with the introduction of digital techniques. Audio starts and ends with analog, since we can only hear in analog. But between those two points, processing is now being handled in digital form. Chapter 6 explains just how this is done.

An in-home music room and a concert hall are at opposite ends of the spectrum, and yet both are an integral part of music production and recording. They are effective, inseparable parts of any solo instrumentalist or extensive orchestra. Chapter 7 introduces the subject of acoustics and explains how to exercise control over acoustics.

At one time musicians, whether amateurs or professionals, were likely to be limited in their expertise to one or possibly two instruments. But now, with the help of electronic synthesizers, even the musical hobbyist can produce orchestral sound. The formerly solo instrumentalist has a choice of dozens of instruments. He or she has been given the ability to orchestrate them, to put the sound into an electronic memory for recall at some later time, and to produce sounds that have never been heard before. In short, the amateur has been given a hand up into a professional class, whereas the professional can now exceed his or her own capabilities. The methods and techniques are described in Chapter 8.

Surround sound is not new. It was a natural sequel to the introduction of stereo, but when electronics made it possible, there were no less than three or four competing systems, thoroughly confusing all those who had hoped to make a home listening room the equivalent of a concert hall. Surround sound is back;

using new techniques, it may produce new, enjoyable levels of music reproduction in the movies and in the home. Chapter 9 details the modern systems.

The purpose of the recording of sound is to permit its reproduction as often as wanted. The professional studio is still the best technique, but in-home recording enthusiasts can do a creditable job by following the methods suggested in Chapter 10. Setting up a studio can be time consuming and expensive, but—as explained in this chapter—can lead to satisfying results.

ACKNOWLEDGMENTS

A number of highly cooperative manufacturers supplied substantial amounts of material that helped form the basis for this book. They were kind enough to open their files, to talk to me, and to supply essential data. I would especially like to thank AKG Acoustics, Inc. for the information they gave me on the construction and use of microphones. I am also indebted to Shure HTS for their material on both in-home and theater four-channel sound.

There were other sources and I am indebted to them as well. They include

Audio-Technica U.S. Inc.
Crown International Inc.
Crystal Technology Co.
Electro-Voice Inc.
Electronic Music Laboratories, Inc.
Gould, Inc., Audio Pulse Division
Nakamichi Co.
Onkyo USA Corporation
Proton®
Robert S. Schulein, General Manager, Shure HTS
Sennheiser Electronic Corporation
Sony Corporation of America
Yamaha Corporation

Please accept my thanks and my gratitude.

Martin Clifford

1

Elements of Sound

Awareness of audio extends back in time for about 2,000 years (and possibly longer); the ancient Romans had a word for it, represented by the Latin verb *audire*, "to hear." This simple definition has since been expanded to include the concept of sound as one of the relatively few sources of energy. Other than sound these include chemical energy, electric, heat, potential, kinetic, nuclear, and light energy.

SOUND WAVES

We live at the bottom of a sea of air. If the molecules that comprise air are not disturbed and remain fairly motionless, or are in moderately random motion, they do not produce sound. If numbers of them are made to move, such as when the wind blows, we hear sound. The molecules can be made to move rapidly, as in the case of hands being clapped. Unlike water waves, whose motion is transverse—that is, up and down—sound waves are longitudinal, moving back and forth. A longitudinal wave is one in which the particles of the medium vibrate in the direction of the line of advance of the wave.

Sound is the result of the vibratory motion of some object that supplies its energy to molecules of air. As a result, these molecules assume a to-and-fro motion, a displacement from their random movement. If this motion is regular, it is described as *periodic* and the resulting sound is musical. If the motion is irregular, it is considered *aperiodic* and is classified as noise.

The demarcation between noise and music isn't always sharply defined. In some instances noise may be included as part of a musical score, possibly to supply sound contrast, as a background to music, or for special effects. Quite often, natural noise sounds such as thunder, street traffic, or insect sounds, may be within a narrow frequency range. When the bandwidth is narrow, it assumes the property of sound location by pitch, a characteristic of musical tones.

THE SOUND MEDIUM

For sound to move from one place to another, it must have some medium through which it can travel. That medium can be a gas, liquid, or solid, but for almost all sound transference, the gas medium is air.

Sound Storage

Sound can be perceived and stored in the human brain, and it can be remembered and taken out of that storage. Sound can also be stored on a phono record, in the form of holes on paper tape, as varying discrete magnetic fields on magnetic tape, and as microscopic bumps on a compact disc. But no matter which of these methods is used ultimately, the stored sound must be converted to a vibration of air molecules to be heard. The process is often made possible through the use of transducers, devices that convert one form of energy to another. A microphone is a transducer, converting sound energy to its electrical equivalent. A loudspeaker is a transducer; working in a manner opposite that of a microphone, it changes the electrical equivalent back into sound energy.

Pitch

Pitch is a characteristic of musical tones and is sometimes thought to be synonymous with frequency, but there is a difference. Pitch includes not only frequency but intensity as well. Pitch is directly proportional to frequency: the higher the frequency, the higher the pitch. Due to the nonlinearity of the human ear, variations in pitch will occur with a variation of sound intensity (sound loudness). Loudness is subjective.

Inverse Square Law

Sound is three-dimensional radiation from a source capable of projecting sound energy. The intensity of the sound radiation decreases in proportion to the square of its distance from the sound source.

Acoustical Masking

If sufficiently loud, one sound can cover, or mask, another, making it inaudible. This effect is most pronounced in the mid- to treble range. Masking can be a two-way effect, since noise can mask musical tones or vice versa, or one musical tone can mask another. Masking can be used in noise-reduction systems.

SOUND PRESSURE LEVEL (SPL)

Sound pressure is a reference to a change in air pressure that is above or below a barometric mean. Sound pressure level is a variation above and below normal atmospheric pressure. Atmospheric pressure is not dependent on sound; rather, sound is dependent on the production of changes in atmospheric pressure.

THE TRANSMISSION MEDIUM

One of the most distinguishing characteristics of light, as a form of energy, is its independence of a transmission medium, because it can travel through the vacuum of space. Sound cannot travel through a vacuum. In an electric light bulb, the popping of a filament remains unheard as long as the glass forming the bulb remains unbroken.

For sound, the functioning of the medium is its work as a transmission agent. Sound energy is supplied by a source, that energy is transmitted by a medium (most often air), and then the sound energy is received by and perceived by a listener.

Energy Level of the Sound Source

The amount of acoustic energy produced directly by a sound source is usually extremely small. The music supplied by a violin without its body can barely be heard. Because of its small surface area, a violin string has little effect on the air molecules surrounding it. The immediate environment of that string is augmented by the body of the instrument, and it is the vibration of this body and its greater surface contact with air molecules that we hear. It is the violin body that determines the quality and the intensity of the sound, and that includes the coating (such as varnish) used on the instrument.

The intensity of what is heard—its level and its quality as well—are also determined by the environment, which can vary from a room to a hall, an auditorium or concert hall, or an outdoor amphitheater.

THE PRODUCTION OF SOUND

Unless subjected to some external pressure, the molecules of air are normally at rest or in small random motion. Sound is generated when this motion is displaced, as in the case of a musical instrument that is played or hands that are clapped. These actions result in the transmission of longitudinal pressure waves by air molecules. At higher altitudes sound is transmitted less efficiently because of the reduced amount of air.

Molecular air displacement results in disturbances in barometric pressure, consisting of alternate regions of higher and lower pressure. The disturbance of air around a sound source isn't restricted to a single source. It is possible to have two or more adjacent sources, with molecules of air around each disturbed. Further, these separate sources need not supply equal amounts of energy. Thus, the medium, air in this case, can support a number of independent sound waves produced at the same time.

Although air is most commonly used for the transmission of sound, liquids and metals are more capable of transmitting sound, since they have a higher modulus of elasticity. A modulus, usually expressed numerically, is the degree to which a substance possesses a specific property. Elasticity is the capability of a strained body to recover its size and shape after deformation. The higher the modulus of elasticity of a substance, the more readily it transmits sound energy.

THE AUDIO BAND

The audio band consists of all frequencies between 22.4 Hz and 22.4 kHz, as defined by the International Electrotechnical Commission (IEC), but in common usage it is understood to include waveforms extending from 20 Hz to 20,000 Hz, sometimes referred to as the human audible spectrum. The word *audible*, as used here, is optimistic, since it would be unusual for an individual to be capable of hearing sounds at or near the upper and lower limits of this band.

Infrasonic, Ultrasonic, and Macrosonic Waves

Audio and sound energy are sometimes used synonymously, but more properly audio means audible; audible sound is a small part of the gamut of sound. Sound waves above 0 Hz (or pure DC) to 20 Hz are *infrasonic*, or below the bottom limit of human hearing range, whereas those above 20 kHz are *ultrasonic*, referred to at one time as supersonic. No upper-frequency limit has been established.

Although infrasonic (infrasound) and ultrasonic (ultrasound) waves are humanly inaudible, they do have commercial applications: infrasonic for drilling through rock; ultrasonic for detecting flaws in metal, for tenderizing meat, in gallstone surgery, and for the emulsion of otherwise immiscible liquids. *Macrosonic* (macrosound) is another designation that refers only to sound amplitude.

These high-strength waves are used for cleaning small metal parts, for emulsifying liquids, metal plating, soldering, and drilling.

THE SOUND SPECTRUM

The *sound spectrum* is the frequency range between the lowest and the highest tones to be heard, stored, transmitted, or reproduced. It is also known as the *bandwidth*, and like the electromagnetic spectrum, is specified in hertz. Bandwidth is considered to be all or some selected portion of the sound spectrum. The frequency range of the bandwidth of an electronic circuit must be specified. That of the sound spectrum need not be. Both the bandwidth of a circuit and the sound spectrum are within the audible range and do not include infrasonic and ultrasonic frequencies.

White Sound

White sound is sometimes confused with white noise. White sound consists of the entire audible frequency range and comprises all fundamentals, including odd and even harmonics.

White sound isn't sinusoidal, but is a complex waveform having a steep wavefront and containing all frequencies that can be perceived by the human ear. As its frequency increases, its amplitude decreases.

Black Sound

Black sound consists of any sound either below 20 Hz or above 20 kHz—that is, outside the audio frequency range. However, it is also used to describe sound in the audio range whose pressure level is so low as to be completely inaudible. Inaudible in this case does not mean a complete absence of sound pressure; instead, it simply refers to a lack of ear response. Ear response can vary from one individual to another, depending on ear sensitivity.

THE SINE WAVE

For all sounds, the basic waveform is a sine wave, also called sinusoid (Figure 1-1). It can be produced by a flute played softly or by a tuning fork. A basic sine wave is not produced by musical instruments, but it can be generated electronically.

A pure sine wave is characterized by a positive and a negative half, each of which is the mirror image of the other. Because of this changing polarity, the sinusoid is an AC wave, with its frequency determined by the number of complete cycles generated during each second of time.

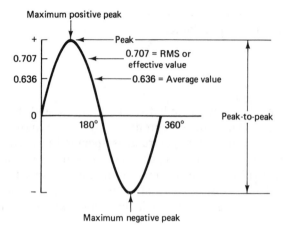

Figure 1-1 Measurement characteristics of a sine wave. The horizontal axis can be in degrees or time units.

A pure sine wave has no harmonics, but the generation of such a wave can be difficult. A resulting distortion, however slight, indicates the presence of harmonic energy. A sine wave is given its name because it varies in proportion to the sine of an angle.

HARMONICS (OVERTONES)

Musicians refer to *harmonics* as *overtones*. Harmonics and overtones are synonymous except in the way in which they are numbered. In using harmonics, the fundamental frequency is known as the first harmonic. The second harmonic has twice the frequency of the fundamental, the third harmonic has three times the frequency, and so on. Musicians do not refer to the fundamental wave as other than the fundamental. A frequency having a value of twice the fundamental is called the first overtone, one having a frequency that is twice the fundamental wave is the second overtone, and so on. A complex waveform consists of a fundamental and a number of harmonics, or overtones.

Harmonic Analysis

Although an audio waveform can appear to have a complex shape, it can be resolved into a number of constituent harmonics, all of which are sinusoids and all of which have an arithmetic relationship to the fundamental wave. Each harmonic can be two, three, four, or more times the frequency of the fundamental. Harmonic distortion, sometimes referred to as amplitude distortion, exists when there is no relationship with the fundamental.

The waveforms of the harmonics produced by musical instruments, no matter what their complexity, can be analyzed and determined by a relationship known as a *Fourier series*.

NONHARMONICS

Harmonics, multiples of fundamental frequencies, are so often associated with musical instruments that there may be an inclination to regard all tones as having such a formation—that is, consisting of a fundamental frequency plus harmonics. However, there are some exceptions. A bell, for example, can produce a sound that consists of many overtones that are not arithmetically related to the fundamental frequency. This is also true for wind instruments such as the trumpet and trombone when these instruments are blown very hard.

Artificial Harmonics

It is possible to produce the harmonic of a tone without, at the same time, producing the fundamental of the tone itself. This can be done on a violin, for example, by very lightly bowing at a point where the second harmonic would normally be played. The resulting sound has a very low level of loudness.

Low Frequencies

Frequencies corresponding to very low tones in the subcontra octave—that is, below about 32 Hz—are used only in rare cases by composers and are rarely played by musicians. Other than through the pipe organ, these frequencies are not considered for composition and reproduction.

MEASUREMENTS OF A WAVE

The positive peak of the wave indicates maximum compression of air molecules; the maximum rarefaction is represented by the negative peak.

Wavelength

The distance from one positive peak to the next or from one negative peak to the next is the wavelength, often indicated by the Greek letter lambda (λ).

Periodicity versus Waveshape

A *periodic wave* is one that repeats itself uniformly in equal amounts of time. Periodicity is applicable not only to a sine wave but also to waves having any shape. A musical waveform can be periodic or aperiodic. Further, there can be an intermix of periodic and aperiodic waveshapes in a musical composition. To be periodic, a wave must meet certain criteria.

Wave Identification

Waves can be categorized as compressional, longitudinal, shear, transverse, plane, spherical, or cylindrical, depending on the shape relationship of the wave to the medium in which it is being transmitted. Thus, a *transverse wave* is one in which the elements of the medium move in a direction that is perpendicular to that of the wave motion; a *spherical wave* is one in which the wavefronts are concentric spheres, a *plane wave* is one in which the wave fronts are parallel planes at right angles to the direction of propagation. Sound waves are longitudinal.

In a *longitudinal wave* the molecules that constitute air are alternately compressed and rarefied, with this action moving away from the sound source in all directions. A longitudinal wave is one in which the air molecules (assuming the sound is traveling through the air) vibrate in the direction of the line of advance of the wave.

Determination of Frequency

Before it can generate a sound, a string made of a substance such as wire or gut must have elasticity. As indicated previously, elasticity is the capability of a stretched material to recover its size and shape after deformation. The frequency of vibration is established by four physical characteristics: length, tension, mass, and density.

Length.   The frequency, measured as vibrations per second, or hertz, is inversely proportional to the length of the vibrating string. Thus an instrumental string that is 10 in. long will have a frequency one-half that of a 5-in. string, assuming they have the same tension.

Mass.   The vibrational frequency of a stretched wire is inversely proportional to its thickness. Thus a wire having twice the thickness of another wire would vibrate at one-half the frequency of the second wire.

SOUND FIELDS

The sound supplied by a source such as a musical instrument or a voice is called a *sound field*; it is often considered in two sections, although there is no discontinuity between them. The field closer to the source is called the *near field*. The other, more removed, is the *free field*.

Near a source of sound, the sound waves travel in the shape of a sphere; these comprise the near field. At a certain distance the radius of this sphere becomes so large that the sound propagation occurs in a parallel manner; this is then referred to as the free field.

The sound pressure decreases proportionally with increasing distance from the sound source, whereas the velocity decreases with the square of the distance in the near field and is directly proportional to distance in the free field.

RESONANCE

Resonance can be mechanical or electrical. Mechanical resonance is the vibration of some object caused by the application of a periodic stimulus having the same or nearly the same period of vibration as the natural vibration period of the object. When a physical object is stimulated by being struck or otherwise set into motion, its maximum vibration will occur at its natural frequency of resonance.

Electrical resonance is attributed to any circuit consisting of an inductance and a capacitor with these connected in series or in parallel. The actual resonant frequency depends directly on the value of the inductance in henrys and the value of capacitance in farads. The circuits are in resonance when their reactances are equal. When either the inductance or the capacitance (or both) is changed, the resonant frequency is also changed. This characteristic is widely used in filters.

Resonance can be characterized as sympathetic or forced. *Sympathetic resonance* can occur in an object whose vibrational frequency is similar to that of a nearby sound source. A classic example is the fracturing of a thin wine glass by the sound of a voice.

Forced resonance does not depend on the similarity of resonance of a sound source and some object. The use of a sounding board by a piano is an example of forced resonance. The transfer of sound energy from a tuning fork to a wooden table top is another example.

Wolf Tones

Efforts are made in the construction of string instruments, such as the violin, viola, and piano, to keep the body of the instrument and the sounding board from having a natural period of vibration within the audio range. A natural resonance in such instruments results in a greatly strengthened tone having a particular frequency; it is referred to as a *wolf tone*.

SONIC EFFICIENCY

Sonic efficiency is the ratio of the sound energy produced by mechanical energy. The decimal resulting from this division, multiplied by 100, is the sonic efficiency.

Sonic efficiency is extremely low. The energy content of ordinary sounds is miniscule compared to the mechanical energy producing it. The volume of sound is no guide to sonic efficiency; a loud sound does not necessarily contain substantially greater energy than one that is soft. To produce sufficient sound to supply 1 horsepower (hp) (1 hp is equivalent to 746 watts) would require a band of 10,000,000 musicians playing fortissimo.

As another example, consider the change of the energy contained in ordinary conversation to heat. If this energy were to be used to warm a cup of tea, it would require continuous, uninterrupted conversation for 90 minutes, by a million people. The resulting sonic power of 10,000 people cheering at the tops of their voices at a political convention changed to its electrical equivalent might be enough to light a single 40-W bulb.

Sonic efficiency is much higher when the source energy is electrical rather than mechanical. As energy converters, loudspeakers are much more efficient as energy transducers than microphones. Microphones are mechanical; loudspeakers are electrical.

PHONONS

Sound is ordinarily considered only in terms of waveforms. This is one way of describing its behavior, but it ignores the fact that the movement of sound from a source to our ears is the repetitive transfer of energy. It takes energy to produce sound and to produce a response in our ear mechanism.

To consider sound from an energy viewpoint, it is described in terms of phonons. A *phonon* is a quantum of sound and is the smallest unit of energy of the vibration of air molecules produced by a sound source. It is comparable to a quantum of light, or a photon.

The phonon may be regarded as a packet of sound energy behaving like a particle that has no mass. This particle is endowed with energy identified as hv, where h is Planck's constant and v is the frequency of the vibration of the sound. The phonon also has momentum indicated as h/λ, where h is Planck's constant and lambda (λ) is the wavelength. The product of the wavelength and the frequency, (λv), is v, or the velocity of the sound.

ARTICULATION

The purpose of verbal communication is to convey thought. Various tests have been designed to determine the number of words generated by a speaker that are determined to be correctly received by various listeners. The ratio of the number of words spoken and correctly comprehended compared to the total spoken is referred to as the percentage of word articulation.

Articulation Index

The articulation index is a measure of the requirements involved in the determination of articulation. These include audibility; the speech must be above the level of audibility—that is, it must be heard to be perceived. Perception of a word depends on the noise level and the amount of masking effect of that noise. There is also an upper level to sound pressure, above which vocal sounds cannot be understood.

The frequency range of the human voice extends from about 200 Hz to 6 kHz. Comprehension lessens for vocal sounds outside this range, and in a closed environment, such as a room, hall, or auditorium, comprehension is also affected by reverberation time. Excessive reverberation can reduce or destroy vocal sound perception. As in the case of music, speech has a dynamic range, which is about 30 dB. The sound pressure level heard at the human ear, including both articulated words and ambient noise level, in decibels is 0.0002 dyne/cm^2.

For sound reinforcement situations, articulation is also dependent on the speaker's knowledge of microphones and their use.

FORMANTS

The sound of an instrument is affected by the way it is constructed and by the material of which it is made. In the case of the violin, as an example, the shape, the materials, and even the varnish used will help determine the sound. Obviously, then, not all violins produce the same sound quality, and this statement is applicable to all other instruments as well.

The factor that makes one instrument in the same family of instruments sound better than another is called a *formant*. It expresses itself by emphasizing certain frequencies. Charts and graphs have been prepared that supply the formants for various instruments, indicating the intensity of formant tones compared to the fundamental.

DYNAMIC RANGE

Dynamic range is the power ratio expressed in decibels (dB) between the strongest and weakest levels of a sound source. It can also be considered the sonic distance between a noise floor, and, at the other extreme, the point at which distortion prevails. As an example, an orchestra can have a dynamic range of 90 dB. The dynamic range of phono records is often compressed. Low levels can be increased to avoid the noise floor, and high levels can be decreased to avoid distortion.

Dynamic Headroom

Dynamic headroom is a much more important amplifier specification than root-mean-square (RMS) power. Although an RMS rating is useful when working with components such as light bulbs and resistors, it has no relevance to music.

Music is never as static as a steady-state test tone. It is full of spikes, peaks, transients, percussives, and so on. RMS power ratings are so ubiquitous only because they were mandated more than a decade ago to curb abuses of nonstandard ratings. Many confusing terms were used then, such as peak power, impulse power, and music power, all designed by the advertising and publicity departments of less-than-ethical manufacturers.

An Institute of High Fidelity (IHF) standard exists for measuring dynamic headroom. This standard is the ratio between the amplifier's short-term power, measured for peaks lasting 20 milliseconds (ms) out of every 500 milliseconds and its rated RMS power. Most amplifiers have less that 1 dB of dynamic headroom, meaning their maximum short-term power is only 25% more than their RMS rating. A few amplifiers even have 3 dB of headroom, or 100% more power (two times as much) for peaks. The aim for having a higher dynamic headroom is not to win a power rating race. Rather, it is to provide consistently heightened realism, more openness, and a more natural sound.

Sustained Power

The IHF's dynamic power test requires an amplifier to deliver its peak dynamic power for only 20 milliseconds (0.02 second). Presumably the IHF thought that peak power output is rarely needed for a longer time than that. Research, however, has demonstrated that this 20 millisecond duration is too short because some dynamic peaks of music, counting the attack and the decay, actually persist for up to 300 milliseconds, 15 times longer than the IHF requirement.

As an example, consider the powerful punch of a bass drum. It doesn't die out in only 0.02 second. Ideally, then, an amplifier needs reserve power for both the short attack transient and for the longer decaying sound. Figure 1-2 illustrates musical dynamics and the necessary power requirement.

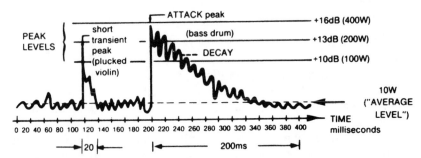

Figure 1-2 Some music peaks, especially those created by synthesizers, have attack peaks 10 to 16 dB above average level and decays in excess of the 20-ms IHF requirements (Courtesy Proton©).

Maximum Dynamic Power

It is desirable for an amplifier to be able to deliver maximum dynamic power for up to 400 milliseconds (Figure 1-3). This ensures that more than enough reserve power will be available for all the dynamic requirements of music from conven-

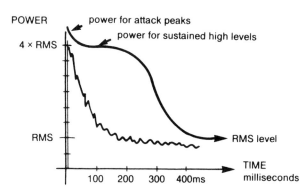

Figure 1-3 Power required for signal with strong attack peak and strong decay (Courtesy Proton©).

tional musical instruments to electronic synthesizers, which can produce longer-than-normal sustained peaks. High-power capability for peaks is a necessity. In uncompressed digital recordings, peaks are typically 10 to 20 times higher than average levels.

Amplifier Efficiency

Efficiency is important because it means less heat will be generated. Since the life of electronic components is shortened by heat, a cooler-running amplifier (high efficiency) will last longer than a hot amplifier (low efficiency).

Most amplifiers fall into these classifications: Class A, B, AB, H, or C. Class H is the best in terms of efficiency, followed by Class B, AB, and A, with A the lowest. Class H offers no benefit over the other designs, since it becomes less efficient when delivering continuous high power. There is also a Class C amplifier, a very high efficiency type, but not suitable for audio, since current flows through it for only a small part of the audio cycle. It is used for CW (continuous wave), or code operation.

Power Dissipation

An appreciation of power dissipation can be had by comparing two widely used amplifiers, Class B and Class H. In both of these a matched pair of power transistors is used to handle both halves of the audio input cycle.

Figure 1-4 shows a typical Class B amplifier. The power wasted (commonly referred to as *dissipated* power) is high, and the efficiency is poor.

Figure 1-5 shows the conventional Class H circuit, with two supply voltages, V_{HI} and V_{L0}. SW_1 and SW_2 are switches controlled by the signal level.

When the peak of the output signal goes higher than V_{HI}, SW1 and SW2 will close. This will enable supply V_{L0} to be sent automatically to the output stages, Q1 and Q2, to prevent clipping. When the signal level falls below V_{L0}, then switches SW1 and SW2 will open and V_{HI} takes over again as the operating voltage.

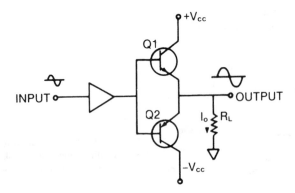

Figure 1-4 Schematic of class B amplifier (Courtesy Proton©).

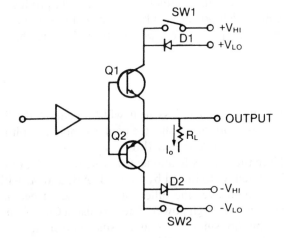

Figure 1-5 Schematic of Class H amplifier (Courtesy Proton©).

In Figure 1-6(a) and (b), the shaded areas indicate that Class H power dissipation remains low, where the efficiency is kept high. However, if the V_{HI} supply were to be used for any extended period—that is, when the input is a sustained high level—the dissipation and efficiency would become much worse, as shown by the dashed line in Figure 1-6. The result could be overheating.

Clipping Headroom

Clipping headroom is the ratio of an amplifier's rated continuous power to its actual power measured at the onset of clipping. If two amplifiers have the same clipping headroom, the one with the higher dynamic headroom rating will produce higher average levels without distortion.

Transients

Transients are peaks, brief bursts in the amplitude of a signal. These can be 10 to 20 dB higher than the average level but are usually too short in time duration to be perceived as being louder. They are, however, essential to musical realism.

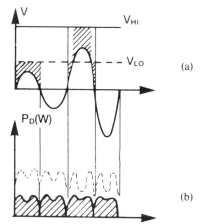

Figure 1-6 (a) Signal voltages for class H with low and high signals; (b) Class H power dissipation for low and high (dashed line) supply voltages (Courtesy Proton©).

Beats

When two pure notes—that is, sine waves not accompanied by harmonics—of identical frequencies are generated simultaneously, the difference between them can be written as

$$f_2 - f_1 = 0$$

where f_1 and f_2 are the two frequencies involved.

If either one of the frequencies remains constant while the other is increased, the difference frequency will move from its zero value to some discrete amount. This can be written as

$$f_2 - f_1 = \Delta f_b$$

where f_2 and f_1 are the original sine waves and Δf_b is their difference. The Greek letter Δ represents "a change of." The resulting beat frequency is f_b.

As the variable (taken as f_2 in this example) decreases in frequency toward f_1, the number of beats per second also decreases and finally falls below the point at which the beats can be heard. On its way to silence, the resultant wave will be heard as a tone having a very slow rise and fall in loudness. Like the two waves from which it results, the beat is a periodic waveform.

Fused Tone Pitch

When Δf_b is close to or at zero, the individual sine waves f_1 and f_2 are no longer perceptible. What is heard is their union, which is known as a fused tone. Our hearing apparatus is unable to recognize two individual tones when their frequencies are very close.

The presence of a beat frequency is an indication that the two pure waves are close to each other in frequency. In effect, one frequency can be said to be *out of tune* with the other. Beat frequencies have a practical application, since they can be used to determine if two instruments or a single instrument and a tuning fork are in tune with each other. Commonly, the note selected for instrument tuning is A above middle C, which has a frequency of 440 Hz.

Note, as indicated in Figure 1-7, that the two original sine waves are partially out of phase with each other. The resultant wave, the beat wave, does not have a constant amplitude, a characteristic of the origin waveforms. Further, its frequency is lower.

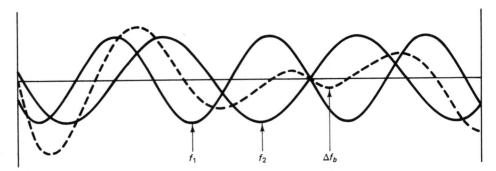

f_1 f_2 Δf_b

Figure 1-7   Formation of a beat frequency.

MUSICAL SCALES

The Chromatic Scale

Musical scales can be described in a number of different ways. A chromatic scale consists of a 12-step octave that includes semitone intervals. A semitone is the tone at a half step. A chromatic scale can be played by starting at middle C on the piano, playing every note using all the black and white keys until the C above middle C has been reached.

The 8-Note Scale

An 8-note scale consists of two scales, a major and a minor scale. Starting at the middle C on the piano and playing all the white notes up to the next C produces a major scale, in this case the scale of C major. The pattern of the C major scale consists of two tones followed by a semitone, which is then followed by another pair of tones, ending with a tone followed by a semitone. Using this arrangement, a C major scale can be played elsewhere on the keyboard.

A minor scale has its own pattern: tone, semitone, tone, tone, tone, semitone, tone.

CHARACTERISTICS OF MUSIC

The notes that form music have a number of different characteristics. They can vary in frequency, in loudness, in time duration, in their resulting waveforms, indicating whether they are percussive or not, and in their energy content. As an example, there is more power in the bass tones than in the midrange and more in the midrange than in the treble. Thus, the energy content of treble tones is about 10 to 20 dB less than that of the bass and midrange frequencies. The energy supplied to all tones is initially via a source such as the voice or an instrument, so the energy demands on that source are much greater for bass tones.

Each tone has a number of harmonics with just a few exceptions and the energy content of those harmonics depends on the amplitude of each, their number, and their frequency.

Human Hearing

The ear is an extremely sensitive hearing device; in terms of sound loudness, it is capable of detecting sounds having a range of 10^{12} from the softest to the loudest sounds. On a ratio basis the loudest sound is a million million times that of the softest sound. The sensations of sound perceived by the brain consist of loudness—that is, the amount by which one sound seems to be stronger than another—and pitch—that is, the frequency of one tone compared to the frequency of another. In terms of tonal energy, the energy of the lowest bass tone on the piano must be a million times (10^6) as great as that of middle C if both tones are to sound equally loud. Middle C has a frequency of 256 Hz or a wavelength of 3.93 m in air.

The Human Ear

The human ear is a barometric transducer, sensitive to changes in air pressure. These changes are extremely small and the movement of the ear drum or tympanic membrane ranges from only 10 millionths to 40 millionths inch.

The dynamic range, the response of the ear to sounds that are minimally audible to those that can cause hearing pain, is more than 120 dB, or a range of a trillion to one. This is greater than the maximum dynamic range of a compact disc player, which is only about 90 to 1. At 1 kHz the threshold of hearing is equivalent to a pressure of approximately 0.0002 dyn/cm^2, referred to as the 0-dB level. At a pressure of 120 dB, using 0 dB as the reference, the pressure is 120 dB, or 200 dyn/cm^2.

The ability of the ear to respond to this extremely wide dynamic range is due to the fact that it does not respond to loudness linearly. Thus, when sound power is doubled, the loudness does not double. Depending on the frequency of the sound, it takes a 10-dB increase in power to produce a doubling in loudness. Although the dynamic range of the ear is 120 dB, its loudness range is 12 to 1.

In effect, what the ear does is to compress a power range of 120 dB into a 12-to-1 loudness range. But although the power range is 120 dB, the human ear can detect a change of as little as 1 dB.

The human ear is not responsive to phase distortion. Phase distortion is due to circuit reactances that have different impedances at different frequencies.

Hearing Gain and Loss

The sense of hearing does not remain constant throughout a lifetime, as indicated in the graph in Figure 1-8. For childhood years, between the ages of 10 to 19, females have a slight hearing improvement in the range from 3520 to 7040 Hz. However, between the ages of 30 to 39, males begin to have a hearing loss extending from 880 Hz to 7040 Hz. During this same time period, females have a slight hearing loss.

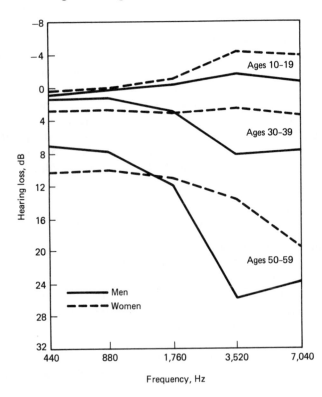

Figure 1-8 Hearing-response characteristics of men and women at different ages.

It is in the age bracket from 50 to 59 that hearing loss becomes most extensive. For males, the hearing loss is about 7 to 26 dB, from 440 Hz to 3520 Hz, after which the hearing ability remains fairly constant; there may even be some small gain in hearing ability. From the age of 10 to the age of 59, females lose a few dB in hearing ability, but this loss remains fairly constant in the hearing range of 440 Hz to 1760 Hz. Following this frequency, the hearing ability gradually slopes until it reaches a loss of about 20 dB in the 50- to 59-year age bracket.

When does human hearing loss begin? In early childhood. Three-month old babies perceive, without difficulty, all tones between 20 Hz and 20 kHz. The upper hearing limit of a man falls by about 2,000 Hz with each decade of life. The 10-year-old can hear up to 18,000 Hz; the 20-year-old, up to 16,000 Hz; the 30-year-old, up to 14,000 Hz; the 40-year-old, up to 12,000 Hz; the 50-year-old, up to 10,000 Hz; and the 60-year-old, up to 8,000 Hz. These, however, are approximations and can vary substantially from one individual to the next.

DISTORTION

Distortion consists of any unwanted change in the waveform or phase of an audio voltage and occurs during the time it moves from its source to its loudspeaker. There are various ways of classifying distortion: frequency, phase, amplitude, and harmonic.

Frequency distortion is caused by the inability of a component to pass what is usually a wide band of audio frequencies. While the human ear is responsive to a range of about 20 Hz to 15 kHz, there is very little music written for frequencies below about 50 Hz. In some instances frequencies below 60 Hz may be cut off to avoid hum voltages. For telephone service, the audio range extends from approximately 250 Hz to 2.8 kHz. This is sufficient for good articulation and is technically more sensible than a wideband system, since it eliminates substantial interference.

A wider audio band is used for AM broadcasting and is kept to 5 kHz. Since the modulation process results in two sidebands, the total required bandwidth is 10 kHz, keeping the signal within its allotted bandwidth. FM broadcasting uses a much wider bandwidth and, as a consequence, suffers less distortion. In these systems, failure to reproduce the bandwidth of the generated signal results in distortion.

Phase distortion is caused by a change in the angular relationship of waveforms to each other and is a shift in this relationship in the output of an amplifier compared to the input. Phase distortion is applicable to the fundamentals of waves and their harmonics. The consequences of phase distortion may be a fluttering or buzzing sound. Phase distortion sometimes occurs due to a malfunctioning phase-locked-loop (PLL) receiver circuit.

Amplitude distortion is the result of a signal saturation of some component. This can be due to the top and/or bottom clipping and can be caused by excessive signal strength. Since the output waveform no longer follows the input faithfully, the result is the production of harmonics that have no relationship to the fundamental frequency.

Harmonic distortion can not only include phase distortion, but also a change in the shape of a harmonic (or harmonics) compared to the original waveform shape. It is also possible for an output waveform to have one or more attenuated or excessively amplified harmonics.

TRANSIENTS

A transient consists of a wave of voltage or current having a sharp peak that lasts for a very short time. Various percussion instruments, such as the piano, banjo, guitar, and mandolin, are capable of producing numerous transients.

In the formation of a sound wave, more energy is supplied at the start of the wave than during its sustain time or at any other part of the waveform. The shape of the wave produced at its start is called a *transient*.

Transient Response

When a square wave is fed to the input of a power amplifier, the result indicates the amount of time it takes for the vertical side of the waveform to reach its maximum value and is measured in microseconds. The specification for this test should indicate both rise and fall times—that is, the amount of time it takes the start of the wave to reach its peak and the time it takes for the peak to drop to zero. An ideal wave would show zero rise and fall times, but in practice 2.5 microseconds (2.5 μs) is considered excellent.

BASIC CHARACTERISTICS OF SOUND

Musical sounds have certain characteristics in common to a greater or lesser degree. We can assume that such sounds start at a zero level and then rise to some maximum value. The difference between a pair of sounds is in the length of time it takes the sounds to reach a peak amount. The amount of time it takes to go from zero to maximum is called *attack time*.

The sound may then remain for a while at some level, possibly a little below maximum, as shown in Figure 1-9. This time period is known as the *sustained sound level*, and, of course, it will vary from one sound to the next.

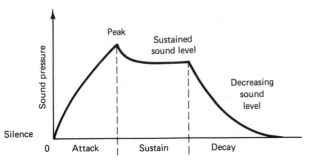

Figure 1-9 Basic characteristics of a sound wave: attack, sustain, and decay.

Finally, the sound will decrease to the zero, or silence, level. This decrease, known as the *decay time* of the sound, may be represented by a more-or-less gradual slope.

Figure 1-10 shows some of the possible variations in the three sound characteristics attack, sustain, and decay. Part (a) is that of a sharp attack and a gradual decay. Note there is no sustain time. Part (b) illustrates a gradual attack and a sharp decay. Again, there is no sustain time. Finally, in part (c) the sound has a sharp attack, a short but sharp partial decay, followed by sustain and then a gradual decay. A large number of partial arrangements can be obtained, depending on how an instrument is played. Using the sustain pedal of a piano, for example, can result in a protracted sustain level, possibly followed by an extremely sharp decay. Since the piano is a percussive instrument, its rise time (attack) can be extremely sharp. A guitar can have a long decay time.

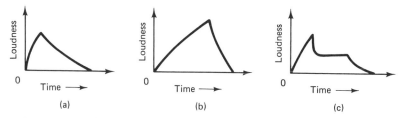

Figure 1-10 (a) Sharp attack, gradual decay, no sustain time; (b) gradual attack, short decay, no sustain sound level; (c) sharp attack followed by a composite sustain. First part of sustain is sharp, followed by gradual sustain. Decay time is short but is also gradual.

SPATIAL HEARING

When sound comes directly from the left or directly from the right, it has to cover a distance of approximately 13 to 16 cm to reach the ear on the other side of the head. Sound travels 330 meters (m) in one second. Consequently, less than 0.001 second is required for the sound to reach the other ear, and this is the longest time delay.

In cases of a sound coming only slightly from the right or left, the time difference shrinks to the order of less than 0.0001 second. However, our hearing is still able to evaluate exactly the time delay difference of such a small quantity.

Since our hearing cannot be switched off, we hear everything spatially, including the range from traffic noise to the sounds of a symphony orchestra.

TIMBRE

Timbre could be called the tone color of each musical instrument, giving each instrument its particular identity. Timbre is based on the number and amplitude of the harmonics generated. It distinguishes a violin, characterized by a high number of harmonics, from a flute, which has relatively few harmonics of low amplitude. Some instruments generate both odd and even harmonics; others, such as the clarinet, supply odd-numbered harmonics only.

Tone Color

Tone color is a characteristic of musical tones that enables us to distinguish one kind of sound source from another. A note played on a violin and a note of the same frequency played on a piano are readily differentiated. The characteristic that lets us make this differentiation is determined by the harmonics of the fundamental tone produced by each instrument.

If the harmonics of a specific piano tone could be made identical to the harmonics of the same tone produced by a violin, that piano tone would sound as if it were generated by a violin.

Figure 1-11 is the graph of a fundamental tone and its harmonics for a piano and a violin. The fundamental tone is that of the note A played on a piano with the same note played on a violin and is 440 Hz in both cases. The amplitudes of both fundamentals are identical, but the harmonics are quite different. If the maximum amplitude points were all connected, the result would be the waveform of the tone. Such a waveform would demonstrate how different the two individual waveforms are.

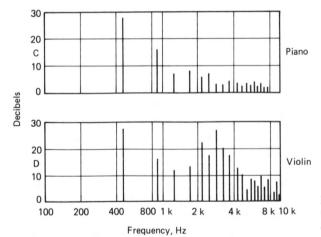

Figure 1-11 Instantaneous values of the fundamentals and harmonics of a 440-Hz tone of a piano and violin.

A fundamental and its harmonics are all basic waves—that is, sine waves—as shown in Figure 1-12. However, when all these waves are combined, the result is a complex waveform. The fundamental supplies the tone with its pitch, whereas the harmonics give the tone its color. Note that the frequencies of the fundamental and of the combined waveforms are the same.

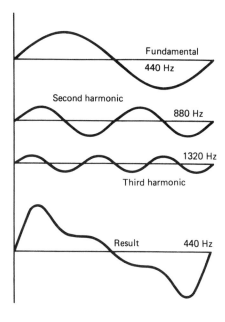

Figure 1-12 Sine-wave fundamental and its sine-wave second and third harmonics. The resultant wave is not a sinusoid.

Concords and Discords

When two notes having a whole-number ratio to each other are played simultaneously, the result is a pleasing sound, alternatively called a *consonance*, or concord. If the notes do not have a whole-number ratio, the sound can be less pleasing and is called a *dissonance*, or discord. This does not mean that discordant tones are to be avoided. They are sometimes chosen deliberately, supplying a variation from consonant tones or as a background.

LOUDNESS

The loudness of a sound depends on two factors: the energy content of the sound and the sensitivity of the listening apparatus—that is, our ears. The frequency response of our ears is a variable; consequently, bass tones appear to be quieter than midrange sounds having the same intensity level. This effect is most noticeable when overall sound levels are low and can be observed when the gain of an amplifier is reduced. Rotating a volume control counterclockwise reduces the bass response much more than the midrange. Compensation is supplied in audio amplifiers by the inclusion of a loudness control. Its function is to supply some bass boost at low listening levels.

The Fletcher-Munson Curve

Human hearing does not have a so-called flat frequency response. More audio energy is needed for low and high audio frequencies to produce the same sensation of loudness as for midrange tones. The Fletcher-Munson graph in Figure 1-13 shows how the ear responds to various audio frequencies at different sound intensity levels. For very loud sounds, in the 100-dB to 130-dB level of intensity, the sensation of loudness is relatively uniform.

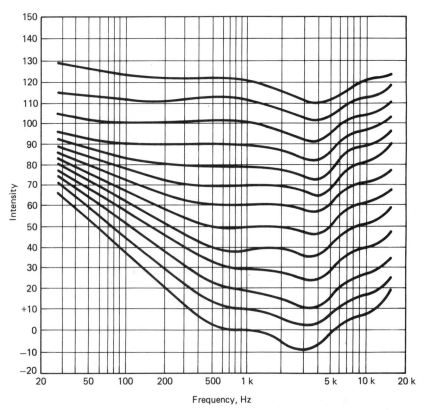

Figure 1-13   Response of the ear to different audio frequencies at various sound levels.

Weber-Fechner Law

Our sense of hearing does not let us perceive sound in direct proportion to its intensity; rather, we perceive it in proportion to its logarithm; this phenomenon is known as the Weber-Fechner law.

Energy Content in the Audio Range

Different amounts of energy are required for the production of musical tones, with the required energy dependent on the loudness required and on the frequency. Bass tones demand the greatest energy input, with the energy requirement decreasing as the frequency moves up through the midrange and then through the treble range. However, a low-amplitude midrange tone can demand less energy than a required high-sound-level treble tone. It takes more energy input to get sound out of a bass fiddle than to play a flute.

$$\boxed{2}$$

Electrical Noise

WHAT IS NOISE?

It is difficult to supply a single, rigorous definition of noise that would be immune from adverse criticism. A more acceptable arrangement would be to supply a number of definitions in the hope that their sum would be a shield against objections.

Noise is a sound that could be called musically unpleasant, any undesired sound or sounds that are not harmonically related, unwanted signals, or static. Whether a sound is noise is subjective. Some sounds consist of dissonances—noise to some, music to others. Sounds that interfere with accepted musical compositions are called noise, as are hum, scratching, clicks, and pops. Noise is widely prevalent, so much so that it is difficult to think of living in a noise-free world. We are so habituated to noise that we accept it as a natural part of our environment.

Unwanted versus Wanted Noise

Some noise is soporific and is highly acceptable; some noise is nerve shattering. Sounds that interfere with music listening—with the recording or playback of music—are called noise. Street sounds, sounds produced by a fuzz box in a musical group, or sounds of surf or a waterfall may be regarded as noise or not. Unwanted noise may include the sound of machinery, a baby crying in the middle of the night, traffic noise, airplane noise, acoustic feedback, background noise, or noise by an audience during a concert.

THE LANGUAGE OF NOISE

Noise can be specific or general. A specific noise is one that pertains to a particular component or some circuit or part associated with that component. Thus, tape noise is specific, and so are microphone noise and quantization noise. Thermal noise is general and can be present in any number of components in which there is current flow through resistors or impedances. Masking effect is generally applicable. Specific and general noises are not restrictive—that is, both can coexist. As a form of energy, noise has its own measurements, characteristics, and descriptive terms.

Some of these terms remain permanently; others become obsolete in time. Thus, microphonics was once applied to vacuum tubes and was noise caused by the vibration of their elements. The language of noise includes references to both specific and general types.

Signal-to-Noise Ratio

Generally abbreviated as S/N, the *signal-to-noise ratio* is a comparison, or ratio, of the amplitude of a signal and that of a noise voltage. Usually expressed in decibels, this measurement is used to make a signal-to-noise comparison supplied by some component such as a preamp, a tuner, or a receiver or some circuit in that component.

Signal-to-noise ratios are poorer for FM stereo than for mono. For mobile equipment S/N is not a fixed amount but is variable, dependent on the distance from a transmitting station and conditions of terrain between that station and the receiver. In the home S/N is also variable and is based on the output power of the transmitter, the distance from the transmitter, the type of antenna pickup, the sensitivity of the tuner or receiver, and the number and type of noise sources in the vicinity.

For signal-to-noise ratios, higher numbers are always better. Table 2-1 is a listing of signal-to-noise ratios versus decibels.

In some instances, as in the case of magnetic tape, the term S/N is synonymous with dynamic range. For tape, the S/N is between an upper extreme of the allowable lower limit of saturation distortion and the tape's hiss, or noise floor. The term *dynamic range* is used more commonly than S/N in this instance.

TABLE 2-1 SIGNAL-TO-NOISE RATIOS
VERSUS DECIBELS

S/N	Decibels
3.2	10
10.0	20
32.0	30
100.0	40
316.0	50
1,000.0	60
3,162.0	70
10,000.0	80

Noise Ratio

Not to be confused with S/N, the *noise ratio* is a comparison between the noise power at the output of a component and that at its input.

Noise Improvement Factor

Abbreviated NIF, the *noise improvement factor* is a comparison of the S/N at the output of some component, such as a receiver, to the S/N at its input. Do not confuse it with noise ratio.

Weighted Noise

Signal-to-noise ratios are sometimes supplied in terms of a *weighted curve*. This technique is used because the weighting network takes the hearing characteristics of the human ear into consideration.

Noise voltages can be sent through a weighting filter and then measured. A weighting filter curve is shown in Figure 2-1. The purpose here is to make noise comparisons more meaningful, not only for a more correct comparison of noise levels but because the noise filter output corresponds more nearly to what the listener hears.

Human hearing is nonlinear and is dependent on perceived sound volume. When the sound has a low level, it has low- and high-end rolloffs. Noise voltages can be generated and sent through a weighting filter. A weighted noise curve helps make more nearly accurate noise comparisons.

Electrical Noise

Electrical noise is similar in its characteristics to an AM signal. A tuner or receiver set for AM reception is often unable to distinguish between AM broadcast signals and electrical noise. Electrical noise is produced by fluorescent lamps, motors, machinery, and automobile engines. It is also generated by electrical storms.

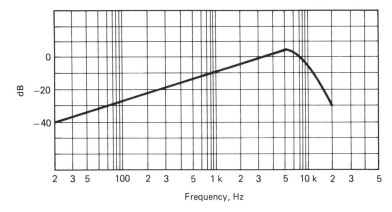

Figure 2-1 Weighting filter curve.

Apparent Noise Level

The human ear doesn't have a linear loudness response and is more sensitive to midrange sounds, so noise in that range has a higher *apparent loudness* than a comparable level of noise at or near the extremes of the audio frequency response.

The average home has a noise level of about 40 dB, an office may have about 55 dB, and a factory may have a noise level of 80 dB or more.

Impedance versus Noise

The random movement of electrons is current flow; this current (I) moves through the impedance (Z) of the conducting substance. This produces a noise voltage, as indicated:

$$E_{\text{noise}} = I \times Z$$

where E_{noise} is the developed noise voltage, I is the current flow in amperes, and Z is the impedance in ohms.

Based on this formula, there are two obvious ways of reducing the noise voltage—by lowering either the current or the impedance. From a practical viewpoint it may be easier to reduce the impedance. But in this case, as in many others, there is a trade-off. In the case of a microphone, for example, the use of a higher impedance means a greater signal-output level. However, the signal-to-noise ratio of both low-impedance and high-impedance microphones remains the same.

The Noise Band

Noise may extend over a narrow or a wide band and in some cases may consist of a series of noise discontinuities. These may have a number of high-amplitude spikes and can produce either a narrow band or a spot frequency punching effect, resulting in dropouts over the musical range.

Noise Analysis

A study of a particular band of noise will include a determination of its frequency components. It may involve the use of instruments such as a noise analyzer for determining the characteristics of a noise.

Noise Blanker

Some receivers, such as the squelch types used for CB, are characterized by high noise levels in the absence of signal pickup. Such receivers have noise-blanker circuitry prior to the demodulator for minimizing ignition-noise pickup.

Clippers

One of the features of an FM receiver is that the intermediate frequency (IF) circuits in the receiver may be equipped with one or more *clipper circuits*, also known as *limiters*, for the removal of noise voltages. These circuits clip the top and/or bottom from signals going through the IF stages (Figure 2-2).

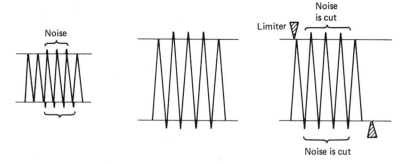

Figure 2-2 Method of noise elimination using a clipper circuit in an FM receiver.

Since the audio information is contained only in the frequency changes, not in the strength or amplitude of the signal, clipping the top and bottom of the IF is a way of removing any amplitude modulation (AM). But AM is exactly what electrical noise is, and so the effect is to remove any noise that may have accompanied the FM signal. Clippers work when a signal of sufficient strength moves through the IF, so noise limiting does not exist alone; rather it is related to the sensitivity of the FM tuner or receiver and the amount of signal pickup by the antenna.

Noise is a random assortment of continuously changing sounds scattered randomly over the entire audio band. However, noise can be supplied with a certain amount of pitch by sending it through a filter.

GENERATION OF NOISE

There are many more sources of noise than of music or speech. Noise is sub-stantially more widespread, more persistent, more continuous, and more prev-alent than sounds that are wanted or enjoyable. Noise is a form of sound energy and so can be generated mechanically or electrically.

Impact, or Mechanical, Noise

Noise produced by mechanical contact between two surfaces is referred to as *impact noise*. Footsteps, the slamming of a door, and the sound of a tire against a pavement are all examples of impact noise. But included in this very small listing is the rubbing of a phono stylus in record grooves. A volume or a tone control, when defective, can produce noise. So can a switch, whether defective or not.

Electrical and Electronic Noise Sources

Microphones produce noise. So do preamplifiers, tuners, receivers, transistors, chips that form integrated circuits, and power supplies. Some devices, such as motors and generators, produce both mechanical and electrical noise. Noise is so important in electronic components that a commonly used spec to determine component quality is its signal-to-noise ratio. The flow of current through a wire can be a noise producer simply by virtue of its movement or by inducing an unwanted noise voltage into an adjacent conductor, either in another wire or other type of metallic element.

Thermal Noise

Thermal noise can be produced by an increase in heat; conversely, it can be reduced by lowering temperature. Thermal noise is the result of molecular agi-tation. When an object is heated, the increased movement of its constituent mol-ecules produces noise. The effect is most noticeable in metals. Not only molecules but free electrons as well are stimulated into greater activity by the application of heat with noise as a byproduct.

Johnson Noise

Johnson noise is also known as thermal noise; it is the noise generated by a current flowing through a resistor at any temperature above absolute zero. The noise power, in watts, can be calculated from:

$$N = KTB$$

where N is the noise power in watts; K is Boltzmann's constant (1.38047×10^{-23}); T is the absolute temperature in kelvins, and B is the bandwidth in hertz. Johnson noise is characterized by a sound that resembles sizzling or frying and is caused by thermal agitation in amplifier circuits, usually in the front-end input of an amplifier.

The Effect of Silence on Noise

Noise is so common and pervasive that we may become aware of it only after it has disappeared or has been somewhat muted. Listening to noise produces fatigue, but since we are often not conscious of it, it is sometimes difficult to assign a reason for continuing tiredness.

Masking

When two sounds are presented to our ears, the stronger may be so dominant that the weaker will not be consciously heard. In the presence of a substantial musical sound, low-level hiss or hum may not be heard, but it will come to our attention as soon as the musical sound is cut off. If the noise level is constant, our perception of it will vary, depending on the momentary amplitude of musical sounds. Hi fi afficionados often play music at high levels, possible responding to a subconscious wish to eliminate noise. A better treatment would be to locate the noise source and to minimize or dispose of it, but this is sometimes difficult or impossible.

Noise Levels

Because the human ear is more sensitive to sounds in the midrange region noise in this part of the spectrum may have an apparently higher level than an equivalent noise level near the extremes of audio range. The levels of noise produced by a noise source may also vary over the audio range. Noise can be not only fatiguing but irritating as well. It impairs the intelligibility of speech and competes with music for our listening attention.

On the plus side, noise can be soporific. Some noise sounds—the wind moving tree branches, the movement of ocean surf, sounds made by some birds, the patter of rain on a roof—can be soothing and, in some instances, sleep-inducing. In making a recording background, audience noise is sometimes deliberately included to supply an air of realism or is followed by a short interval of complete silence to convey the idea of an orchestra about to perform.

Except in an anechoic chamber, getting rid of noise in a live situation is practically impossible. There are so many diverse sources that total noise reduction is out of the question. The best that can be done is to achieve some reduction.

Noise and Recording

During the recording process, noise can be considered as any sound that is not part of the wanted signal. It can be caused by a variety of noise sources—those produced by the environment, an audience, or generated by electronic components, such as microphones.

NOISE IDENTIFICATION

There are a number of different types of noise, which are identified in several ways: by the name of an individual involved in learning the specific cause of the noise, by its characteristics, by a color (such as pink, white or azure), or by its appearance in some part of an audio chain.

White Noise

The spectral distribution of noise is measured in various ways. One of these, *white noise* (also known as Gaussian, or flat, noise)—a form of random noise, as indicated in Figure 2-3—is noise having a constant amplitude, measured in decibels, over a frequency range of 20 Hz to 20 kHz. Its energy content is the same for each cycle of its frequency range.

White noise obtains its name from a comparison with white light, which consists of a mixture of various colors. The development of white noise is from a sawtooth waveform that can be produced by a fundamental sine wave plus odd and even harmonics. If the fundamental is nonperiodic, the resultant output is random and is known as white noise. White noise can be segmented because it can exist over some limited portion of its range, either singly or in bands.

White noise has various practical applications. It is used in musical synthesizers, in most electronic organs and drums, and also in sleep therapy.

Pink Noise

The spectrum for *pink noise*, as shown in Figure 2-3, is the same as that for white noise, extending from 20 Hz to 20 kHz. Unlike white noise, pink noise (a type

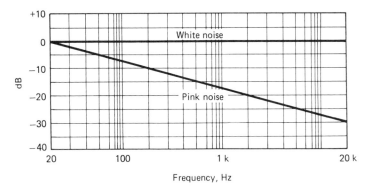

Figure 2-3 Frequency characteristics of white noise and pink noise.

of random noise) has a slope of 3 dB/octave—that is, it falls off 3 dB each time the frequency is doubled. With pink noise the amplitude is inversely proportional to frequency.

The concepts of white noise and pink noise are obtained from a comparison with white light. White light consists of all the colors in the color spectrum. In that spectrum, red is down at the low-frequency end. Pink is a mixture of red and white.

Pink noise can be produced by a generator designed for this purpose and is obtained by passing the white noise output through a low-pass filter having a slope of 3 dB/octave, thus supplying pink noise with its amplitude. Its sound output is used for measuring listening response, stereo response, stereo channel balance, speaker and microphone phasing.

Pink noise is characterized by having a decreasing amount of acoustic energy as it approaches its high-frequency end, and so in this respect is more like music. This makes it more suitable for testing components such as tweeters. White noise, because of its constant energy-level characteristic, could result in damage to tweeters.

Aside from the 3-dB rolloff for pink noise, there are certain important differences between white and pink noise. White noise has equal amounts of sound energy per unit bandwidth. Thus, white noise has the same amount of sound energy between 500 Hz and 600 Hz as it does between 900 Hz and 1 kHz. Pink noise has equal amounts of sound energy on an octave basis. Assume an octave starting at 100 Hz. Since an octave is selected by a 2-to-1 frequency ratio, the second octave would be 200 Hz. The next octave would then be 400 Hz, and the following octave would be 800 Hz. With pink noise, there is equal sound energy between each of these octaves. Further, because of the roll-off of pink noise, white noise is regarded as being higher pitched. Pink noise can also be considered as white noise that is weighted as the bass end.

Adjustments using a pink-noise generator are more accurate and more easily perceived. With pink noise it is possible to recognize frequency dips over the audio range in a listening room. If a pink-noise generator isn't available, it is possible to get some conception of the sound by listening to in-between station noise on an FM receiver.

Azure Noise

Azure noise is white noise weighted at the treble end.

NOISE HABITUATION

It is often easier to control the periodic waveforms associated with music than any accompanying noise. Although noise is frequently undesirable, particularly when it interferes with the enjoyment of music, it is possible to become accustomed to it and to miss it when it is not present, as in the case of phono records. The music supplied by records is accompanied by the noise of the prior tape-recording process. When listening to a phono record, it is possible to hear the noise floor; this becomes evident when the stylus rides an unmodulated groove. This noise becomes inaudible when the groove is modulated because of the masking effect of the music.

Dither

It is possible to become so habituated to noise on phono records that to some, playing noise-free compact discs can be disturbing. As a consequence, digital audio engineers have added random noise to CD masters, using a technique known as *dither*, a low-level noise signal. The effect of dither is comparable to white noise. Not all manufacturers use dither, since this seems to work to the disadvantage of the CD player.

Shot Effect

Solid-state devices such as diodes and transistors are responsible for a type of noise known as *shot effect*. This is noise caused by the random arrival of electrons at the terminals of the diodes or transistors. Even GAs FETs (gallium arsenide field-effect transistors) produce some electrical noise because of the random arrival of electrons at the drain, the output terminal of the unit. However, modern design of GAs FETs is such that the noise they produce is extremely low compared with other transistor types, and for this reason they are widely used in LNAs (low-noise amplifiers) associated with satellite TV systems.

Electrical noise can be regarded as a signal. If a noise voltage is delivered to an amplifier, it will be strengthened just as a signal would be amplified. This will be straightforward amplification of the noise unless the amplifier is equipped with some means of reducing the signal level of the noise or in some way discriminating against it. Such circuitry is found in CB and FM broadcast receivers.

Noise and Bandwidth

Circuits that have a wide band-pass must be capable of handling a wide range of frequencies and are more subject to noise than narrow-band-pass circuitry for several reasons. A wide-band-pass circuit encompasses more of the noise range. Further, wide-band-pass circuits are often characterized by low gain.

Noise Level versus Signal Level

One of the effects of noise is that it places a limit on the smallness of the amplitude of a wanted input signal. Under low-signal-input conditions with the presence of a noise signal of equal amplitude, it may become impossible to separate or determine the wanted signal from the noise signal in the output. In effect, the wanted signal becomes contaminated by the noise to such an extent that the desired signal becomes useless.

The Generation of Noise

All components involved in the process of manipulating an audio signal, whether those components are on-line permanently or intermittently, add noise to the signal. If the signal is being recorded, that noise is added to and becomes part of the signal. Whether that noise is tolerable or not depends on the signal function. For TV a moderate to high noise level may be acceptable; that same amount of noise in a hi-fi system would not be, where a signal-to-noise ratio of even 60 dB could be rejected.

With tape the amount of noise is determined by the physical and chemical characteristics of the particles. The amount of noise is also dependent on the area of tape exposed to the recording and playback heads. The area is determined by the tape width and tape speed. The greater the area, the better the signal-to-noise ratio.

Aliasing Noise

In a compact disc player, *aliasing* is the development of beat note products due to heterodyning between the sampling frequency and ultrasonic audio harmonics. It is possible for the difference frequency, consisting of the sampling frequency minus audio signal harmonics, to fall within the audio range. One of the advantages of oversampling, whether double or quadruple, is the much smaller likelihood of aliasing. Low-pass filters are used in CD players for the specific purpose of eliminating aliasing noise.

Aliasing is referred to as intermodulation distortion (IM) but is generally not included in the specs of a CD player.

Quantization Noise

It was not until the advent of purely digital time-delay techniques using pulse-code modulation (PCM) that an S/N ratio of 90 dB could be achieved. Today, digital delay can meet the stringent requirements of quality studio equipment.

Time quantization is achieved by sampling the signal periodically. A signal is completely defined by its samples if the sampling rate is at least twice the signal bandwidth.

Amplitude quantization results from the conversion of instantaneous amplitudes into binary numbers, causing so-called roundoff, or quantization, noise. To cover a wide dynamic range with a limited number of digits, most digital time-delay techniques use a floating-point format. Above a certain signal level, the floating format produces an almost-constant signal-to-noise ratio with the disadvantage of modulating the quantization noise. To reduce high-frequency noise modulation caused by low-frequency signals, the latter unable to mask the former, the signal is preemphasized at the input of the delay and deemphasized at the output, with the effect that the de-emphasis suppresses the unmasked modulated high-frequency noise.

Noise signals can have a variety of wave shapes. Some of these waveforms are complex, consisting of a fundamental plus a number of harmonics. It is also possible for a noise to be purely sinusoidal. Noise can be direct and reverberant. It can be generated by the mechanical action of musical instruments. To some listeners this adds to natural sound ambience; to others it is distracting.

The goal is not a completely noise-free environment. Rather, it is an environment in which extraneous sounds can be controlled. Controlled does not mean eliminated; rather in the competitive world of sound music should be given a chance at ascendency.

Audience noise—including the jingling of keys, male and female speaking voices, and clapping of hands and footsteps—covers the same frequency range of many musical instruments, with the possible exception of the piano. This is shown in Figure 2-4. If the music level is substantially greater than the noise, the noise will be masked. Music, though, can have a wide dynamic range, and it is during low-sound-level musical passages that the noise will be most evident.

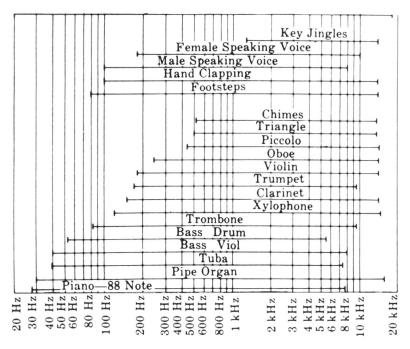

Figure 2-4 Some intrusive noises come within the frequency range of musical instruments.

Effect of Noise on Dynamic Range

Dynamic range is the amount of sonic separation, measured in decibels, from the softest to the loudest sound of a music source. An acoustic guitar will have a smaller dynamic range than a trumpet, another way of saying that the trumpet can produce a much louder sound. For an ensemble or an orchestra, the dynamic range can be the amount of sound level, measured in decibels, from the softest passages to the loudest music it can produce.

Dynamic range is dependent on the noise floor—that is, the maximum amount of noise level. Dynamic range is reduced by this level. If the dynamic range of a signal source is 100 dB and the noise floor is 30 dB, then the dynamic range is the difference between the two, or 70 dB. Reducing noise, then, is a way of increasing dynamic range. For a symphony orchestra the dynamic range is from about 20 dB to 100 dB, with the noise floor affected mostly by audience noise.

Power Line Noise

Power line noise can exist during recording or playback. If a recording group includes an electric guitar, it is possible for line noise to appear in the output of its amplifier. Line noise exists in the form of a voltage. True line voltage exists as a sinusoidal carrier, with the noise voltage as momentary spikes or as a complex continuous waveform. It can also be a 50-Hz or 60-Hz hum voltage fed back into the line from some other connected electric component.

Noise voltages can appear in the AC power line due to several factors. The noise can be fed back to the line from devices that are triggered on and off, such as the cycling of a washing machine, or any other device that has intermittent motor operation. Noise voltages can also be induced into the power line due to signal radiation. The sparking of motor brushes can produce this kind of noise. AM receivers are more susceptible to this kind of noise pickup than FM.

Line noise can be reduced or eliminated by filters. The simplest type consists of a single capacitor placed across the power line connection of a suspected component. The capacitor is placed in shunt with the power line, with one lead wired to the black (hot) wire of the AC line; the other lead is wired to the white neutral (cold) wire of the line.

GATING

A microphone can be "live" even though it may not be in use. At such a time the ambient noise level may become apparent due to the absence of a wanted sound signal, since there will be no masking effect. One of the earliest solutions to this problem was to turn the microphone off with a pushbutton, a function often assigned to the audio engineer. Modern gating devices use sophisticated circuitry to sense the signal, opening to pass programs and shutting down if the signal is below a presettable level. Some recording artists gate virtually every recorded track, eliminating bleed over and resulting in very distinct stereo imaging.

Noise Gating (Correlators)

Some noise-reduction devices use several gates, each responsible for a certain portion of the audio frequency band. Theoretically, these gates are separate and open and close independently. This action allows program frequencies to pass through. Noise signals aren't completely eliminated, but they are attenuated. Various amounts of multiple gating can be used, but it does require excellent and carefully controlled circuitry to be precisely responsive to circuit material. There are various versions of this noise-reducing technique, all of them known as *correlators*.

Gating is usually done while a program is being recorded. It is excellent for keeping extraneous sounds from being recorded, but it is not meant as a cure for other types of noise. As with any device designed to eliminate noise, you must be careful that the solution helps more than it hurts. Well-designed gates will be auditorially "invisible" in operation; they will not chop off the music too soon or let you hear them pop in and out.

Auto-Correlators

Components known as auto-correlators break the frequency spectrum into bands and treat each band individually. An electronic gate in each band opens to let music through and shuts when the music stops to prevent the passage of noise when there is no music in that particular frequency band.

The problem with using noise filters and auto-correlators is that at certain levels the sensing circuits of these devices are unable to distinguish, electronically, between the music and the noise. Thus, these types of frequency-manipulative noise-reduction systems, even when properly adjusted and used, reduce vital harmonic content in the music and often have irritating audible side effects.

NOISE-REDUCTION SYSTEMS

There are various methods used for the reduction of noise. If the recording chain includes microphones, the technique starts with that component. Cardioid microphones (mics) are used to minimize off-axis pickup. With this method the mics are less sensitive to off-axis sound pickup, most notably from the 180° point on that axis. Further, the acoustics of the studio may be altered in order to minimize or prevent noise pickup from external mechanical sources such as traffic sounds, air-conditioning sounds, or voices.

Noise reduction can also be done electronically. There are basically two classes of electronic noise-reduction systems. These are generally referred to as single-ended and double-ended.

Single-Ended Noise Reduction

Also known as after-the-fact noise reduction, the *single-ended* technique reduces noise by eliminating or suppressing those high frequencies where most noise occurs. Most single-ended systems use some filtering, which reduces high-frequency response.

Expanders

Expanders, also known as dynamic range enhancers, operate over the entire sound spectrum and reduce noise by making loud passages louder and quiet passages quieter. The function of the noise gate expander is to eliminate unwanted noise between notes. As the quiet passages are reduced in level, background noise is suppressed. And, as the loud passages are increased in level, the dynamic range of the music is also increased, resulting in a noise-reduced, dynamically enhanced listening experience.

However, expanders must be carefully selected, since some operate more in one portion of the audio spectrum than another. This alters the original musical balance.

An expander can consist of a single-channel noise gate that is stereo-strappable with a compressor/limiter. The two can be used either as a stereo compressor or a stereo noise gate. A noise gate can be easy to use if equipped with single-slider action for quickly setting the amount of gating. A single control knob allows setting the threshold point. Using both the slider and the knob lets the user change the gating characteristics and the sound of the program by ear. The attenuation range is from 0 to 70 dB, depending on the input level and the slider setting.

Double-Ended Noise Reduction

Also known as a compression-expansion system, the *double-ended* approach to noise reduction uses a two-part system consisting of a compressor at the time of recording and a complementary expander at the time of playback.

The audio signal is compressed after the mic preamp or mixer and prior to tape recording to decrease the dynamic range of the original signal so it can pass through the dynamic range "window" of the recorder. The purpose is to keep the quietest signal above the noise floor and the loudest signal below the level of distortion or tape saturation.

Upon playback, the signal is expanded to re-create the dynamic range of the original program material. For a compander to function properly, the compressor and expander must operate as exact mirror images. To control the compression and expansion operation, each must have a signal detector. There are three types of signal detection: peak, average, and RMS.

Hiss Reducer

A *hiss reducer* eliminates annoying, constant background hiss from audio sources without sacrificing treble overtones and is capable of doing so without prior encoding. It utilizes single-slider set-and-forget programming, offering fast and simple dehissing for audio or video tape, samplers, synthesizers, or even the main left and right outputs of a console at speeds from $1\frac{7}{8}$ to 30 ips (inches per second).

When musical or program high frequencies are present, a dynamic filter opens, allowing these frequencies to pass. The filter is a second-order, minimum-phase design topology and operates from 1.3 to 38 kHz, depending on its setting. When meaningful high frequency is no longer present, the filter's action instantaneously "rolls off," or attenuates the high-frequency energy, thereby eliminating hiss. When program highs return, the filter allows the material to pass with little or no sacrifice in signal integrity.

The unit has a quieting control, which is set to adjust the initial amount of hiss reduction to bring the program out of the background hiss. After that, operation is automatic and depends only on the instantaneous spectral dynamics of the program.

Hiss-Reducer Applications

Hiss-reducer applications include the following.

Instrument hiss. It quiets hiss when noisy guitar amps are being recorded.

Video production. It is used where original sound of dubious quality must be edited and dubbed through successive generations.

Dubbing. When dubbing analog tapes in the studio, the hiss reducer cleans hissy recordings to the point where successful release in digital format is feasible.

Broadcast. For transfers or quieting a noisy program chain, an LED (light-emitting diode) display helps the user monitor the dehissing process. A front-panel input to a high-impedance preamp enables musicians to plug samplers, synthesizers, and drum machines into the hiss reducer with operation over a 0- to -20-dB range. Two units may be strapped for stereo.

Noise and Sound Pressure Level

Noise, like wanted sound, is the result of a change in sound pressure level (SPL). SPL is measured in decibels by

$$SPL = 20 \log_{10} \left(\frac{p_2}{p_1} \right)$$

Here, P_1 is the reference level and is unity in terms of relative sound pressure or 0 dB; P_2 is the output level. However, 0 dB, is total silence and is a mathematical concept only. It is the limit of possibility in an anechoic chamber, for even here the SPL is about 10 dB. It might be thought that a whisper would be practically zero, but actually even at a distance of 4 ft it is 20 dB. In a typical home SPL is 40 dB; a representative recording studio is 20 dB.

Microphone Self-noise

Even though a microphone is in the process of being tested in an anechoic chamber, it will still be responsive to noise; in this instance, however, the noise will be generated internally. How this is done depends on the physical structure of the mic. In the case of dynamic types, noise is the result of electron movement in the voice coil and is due to ambient temperature and electron movement of voice-coil currents.

A mic's internally generated noise can be measured. The technique involves using a peak-measuring instrument and a weighting filter (DIN 45 405) to put the right emphasis on the most disturbing components of the composite noise. The so-called equivalent noise is calculated by utilizing the sensitivity of the mic and relating this value to the threshold of hearing. More commonly, the noise of a mic is contained in the signal-to-noise ratio.

Microphone noise is an aperiodic waveform, usually having a very low voltage level, that is present at the output of a microphone with a complete absence of an input signal. Absence of signal means not only no music or voice input, but also no pickup of ambient noise voltages as well. Thus, microphone noise is that generated by the microphone itself and does not include external noise. Microphone noise is a function of temperature and is present in all types of mics.

A mic is an energy converter, an action that involves the generation of noise. To keep noise from being perceived, its level must be kept low enough so it is inaudible, or else it should be masked by having an adequately high level of wanted sound. At a given acoustic sound pressure, it is desirable for the mic to produce as high an electric output level as possible—that is, to have a high sensitivity.

A first step toward the determination of a microphone's self-noise is to use a variant of the Nyquist formula:

$$N(f) = 4kTR$$

where $N(f)$ is the spectral density of the noise voltage, k is a constant equal to 1.38×10^{-23} joules per Kelvin, R is the resistance being measured, and T is the temperature of that resistance in Kelvins. Although impedance is the term most commonly used with microphones, the output impedance of most microphones, including dynamic, condenser, and ribbon types, is almost purely resistive. Since that impedance is resistive, it is not frequency-responsive and so remains practically constant over the audio frequency range.

Unlike the Celsius and Fahrenheit scales, the Kelvin temperature scale omits the word degrees; and its units are in Kelvins (K). Thus, we write 50 K, not 50°K.

The zero point on the Kelvin temperature scale, also called the absolute scale, is equal to $-273.13°$ on the Celsius scale and $-459.72°$ on the Fahrenheit scale. This zero point is absolute zero and is the point at which there is no molecular motion. It is also the point at which thermal noise ceases to exist.

Dynamic Noise Filtering

A *dynamic noise filter* is a device that decides what is music and what is noise and tailors the bandwidth instantaneously to let only the music through. It constantly analyzes and tracks the program material and gradually rolls off those frequencies not needed for the music.

Dynamic filtering can be set up for any frequency and so can be used to eliminate both high- and low-frequency noises. This technique is versatile and, since there is no encoding or decoding, a dynamic filter will work on any program source. Further, dynamic filters can remove noise from existing program material.

A dynamic noise filter capable of following the periodicity of the high-frequency energy can take over where masking leaves off, following the program closely and attenuating the unmasked noise.

The Program versus Noise

Statistically, the greatest amounts of unmasked noise are located above 1.5 kHz. In music the fundamental frequencies of the instruments have high average densities in the lower bands of the audio spectrum, approximately 1 kHz and below. This higher average energy masks the noise in those bands sufficiently. In the upper frequencies, however, the program energy tends to be more sporadic, so this is the problem area. Since there is less energy per unit time, the masking is only periodic and thus the noise is much more objectionable.

MASKING

We are habituated to masking because it is one of the characteristics of sound and is predicated on our listening mechanism. When two sounds coexist, the louder sound masks the softer one, making it more difficult to hear, especially if the two sounds have the same frequency or some harmonic relationship. If a musical tone has a frequency of 400 Hz, not only this fundamental, or first harmonic, but harmonics such as the second at 800 Hz, the third at 1,200 Hz, and so on will be masked by a noise pulse at 400 Hz.

If the noise has a low pitch it can and will mask a higher-frequency sound, even if both have the same amplitude. However, the musical sound can be heard if its amplitude is sufficiently greater than the lower-frequency noise.

Punching

Noise may not only cover a wide passband but may consist of a series of discontinuities over a given frequency spectrum. These may have a number of high-amplitude spikes and so can produce either a very narrow band or a spot frequency punching effect, producing the effect of dropouts over the musical range.

Dynamic Range of a Tape Recorder

The usable dynamic range of a tape recorder is a measure of the difference between the loudest and quietest passages if music can be recorded on the tape and played back without audible noise or distortion. The usable dynamic range can be illustrated as a window of limited height through which the musical signal passes on its way through the tape recorder.

The height of the window is determined by the location of the distortion ceiling and the noise floor. Any signal that is below the distortion ceiling and above the noise floor will pass through the tape recorder without being audibly changed or degraded.

As the signal level—that is, the signal loudness presented to the head of a tape recorder—is increased, a point is reached at which any additional increase in level overloads the tape, and the result is audible distortion. The point at which 3% harmonic distortion occurs is generally accepted as the maximum recording level. The problem for the recordist is that there is no good way of determining when this ceiling has been reached, short of hearing distortion played back from the tape.

Unfortunately, there is no industry standard to relate the 3% distortion point to the recorder's volume unit (VU) meters or overload indicators. Tape recorders may go into audible distortion at anywhere from 2 dB to 12 dB above 0 VU.

Headroom

Headroom is a critical factor for determining the practical ceiling for distortion-free recording. Headroom is the difference between the average operating level of components such as an amplifier or a tape unit and the maximum signal level that can be handled without undue distortion.

Adequate headroom can be overestimated or underestimated. Overestimating a recorder's headroom can result in (1) tape overload distortion; (2) the loss of the sharp edge of percussive waveforms; (3) smearing of peak transient sounds; and (4) a general muddying of loud passages when many instruments are playing ensemble.

Underestimating the headroom and recording at too low a level results in using less of the recorder's dynamic range and places the recorded program at a level where the quieter portions of the music may be below the noise floor of the recorder.

The Noise Floor

The *noise floor* is the restricting factor on the quiet end of a tape recording and is defined as the level at which the recorded program signal is equal to the total hiss and background noise of a recorder. If a tape with nothing recorded on it is played back, there will be a noticeable sound produced by the tape. This sound is subjectively characterized as tape hiss. This hiss, plus any electronic noise contributed by the tape recorder circuitry, defines the noise floor. The noise floor is the lowest level at which a signal may be recorded on the tape and played back without being covered or masked by noise.

Therefore, the usable dynamic range, or window, of any given tape recorder is the number of dB difference between the audible distortion level on the loud end and the inherent electronic noise level plus tape hiss on the quiet end.

For the very best professional studio recorders, the theoretically attainable dynamic range is about 68 dB at 15 ips tape speed, but this performance is seldom met in practice. A more realistic dynamic range for professional recorders is approximately 60 dB at 15 ips.

Modulation Noise

Modulation noise is caused by the process of modulation, or the superimposition of one signal on another. It is sometimes referred to as the noise behind the signal.

Induced Noise

Induced noise can result from electromagnetic induction from the magnetic field surrounding power line currents or any other conductors carrying such currents. It can also be caused by "impure" power line currents carrying unrelated momentary currents.

Parasitic Noise

Also known as trace noise, *parasitic noise* is caused by traces, the connecting conductors on a printed-circuit board that work as antennas, an unwanted function. Digital circuits are particularly susceptible to this effect. Traces not only act as noise-signal pickup devices but can also pick up and retransmit such signals. The effect is to reduce the signal-to-noise ratio of the component in which trace noise is present.

These noise signals also find their way into the power supply and, since this is an element that is common to all other circuits, can cause the presence of noise in components throughout the unit, including those that are not trace-connected. Trace noise can exist in both analog and digital circuits.

Fluorescent Fixture Noise

Fluorescent bulbs are notorious transmitters of noise. The noise is radiated and picked up by nearby electronic equipment. There are several cures, including moving the fixture a greater distance from the electronic gear or substituting incandescent lamps and their fixtures in place of fluorescent lighting.

Motor Noise

Motors that use brushes and sometimes even those that use slip rings can radiate noise. The greater the amount of sparking between the brushes and a commutator or slip rings, the higher the noise level. Although most of this noise is radiated, some of it can also feed back into the power line. To minimize or eliminate this problem, replace the brushes if they are worn. Clean the commutator or slip rings with scrap canvas. Connect the motor to a different branch power line. Electrical noise is AM and can be generated by machinery or automobile engines as well as by fluorescent lamps and motors.

Preamplifier Noise

There are various sources of noise associated with preamplifiers. One of these is noise pickup by the connecting wires between the input of the preamp and associated signal sources. These wires must always be shielded types, and the shield must be grounded. Grounding is usually done automatically when the wires are connected, but it is important to make sure the preamp is grounded. Keep the connecting wires as short as possible.

Turntable Noise

There are three types of noise associated with turntables: rumble, wow, and flutter. *Rumble* is a low-pitched sound. The amount depends on the preamp connected to the turntable. A high-quality preamp that has excellent low-frequency response may supply more rumble output than a unit that has a greater low-frequency roll-off. The price to be paid is poorer bass response.

Rumble can be produced not only by a playback turntable but also is modulated into every phono record. Rumble figures are supplied in dB and should be preceded by a minus sign. The larger the dB number, the smaller the amount of rumble. The minus sign is an indication of the amount by which the rumble is less than the recorded material.

Wow is a sound variation caused by a low-speed change. *Flutter* is a higher-speed quiver and is particularly noticeable in notes that are sustained. Weighted and unweighted figures are supplied for rumble, flutter, and wow. These can be measured with a wow and flutter meter.

Tape Noise

Tape noise is an inherent property of the random distribution of the metallic particles that make up the magnetic layer of the tape. These particles generate signals in the playback head, are random in frequency, and are also low-level. Because of their low signal level, only frequencies where the ear is most sensitive are audible, resulting in the familiar sound of tape hiss.

The absolute level of tape noise is a function of the formulation of the magnetic material and of the size, shape, distribution, polish, and uniformity of the magnetic particles. This noise is not influenced by the recorder or by the level of a recording and, for all practical purposes, is a property of the tape.

Quantization Noise

The conversion of a sample or instantaneous amplitude value of an analog signal into binary form is called *quantization*. In effect, it is the changing of the decimal value of a number of instantaneous peaks into binary equivalents. However, this conversion may not be precise. The difference between the true decimal value and the binary value is referred to as *quantization noise*. However, by using 16-bit binaries and a high-enough sampling rate, this noise can be kept to an inaudible level. Each bit used during the quantization process contributes about 6 dB to the S/N ratio. For a 14-bit system, this would be $6 \times 14 = 84$ dB and for a 16-bit system, $6 \times 16 = 96$ dB. Although an S/N ratio of 96 dB is preferable, a S/N ratio of 84 dB is considered good.

Quasiperiodic Noise

One of the characteristics of noise is that it is aperiodic. Noise is unpitched, and its sound consists of nonrepeatable combinations of a variety of audio frequencies that are not harmonically related. It has also been described as unwanted sound. Music, however, is often defined in a contrary manner—that is, as being periodic.

These definitions are not precise, since some noises have musical characteristics. The problem of whether a sound is musical or a noise is often subjective. Recordings of noise have been made and sold. A synthesizer may be equipped with a noise generator. The output of an electric guitar may be followed by a wah-wah or a fuzz box. Both of these devices are nonmusical add-ons.

Hum

Hum is a noise due to the pickup of a power line's magnetic field or supplied directly by power line wiring. The line frequency is 60 Hz, so noted for its constancy that it is used for electric clocks.

Although hum is correctly categorized as noise, it is periodic and shares this characteristic with music. Hum can be a problem with audio components that have a high degree of amplification, such as pre- and power amplifiers or devices such as microphones that require following substantial voltage gain.

Birdies

A *birdie* is a type of noise noticeable in FM radio receivers that is caused by interference between two adjacent stations. It can be cured by using a narrower IF bandwidth. Some FM sets are equipped with a selector control for easy switching from wide to narrow IF band-pass.

Pre-emphasis and De-emphasis

For the most part, electrical noise exists in the treble portion of the audio frequency range, but tones in this range do not have as much acoustical energy as lower-frequency tones, such as those in the midrange and bass regions. Because of low audio energy in the treble and a relatively high noise level, the signal-to-noise ratio can be poor. In an FM receiver spec sheet, however, the figures on S/N are attributable to the entire audio range and not just to some selected portion.

To improve the FM receiver's S/N, a technique called *pre-emphasis* is used at the transmitting end, and a similar but inverse process known as *de-emphasis* is used in the receiver.

At the transmitter the process of pre-emphasis consists of increasing the strength of treble tones. This upsets the relative strength of treble to bass and midrange tones, a condition that is corrected in the receiver by a de-emphasis network. In the FM tuner or receiver, the treble tones are weakened so the ratio of treble to midrange and bass is correct once again. The advantage is that as the strength of the treble is reduced, so is much of the noise in that portion of the audio spectrum.

Hum and Grounding

All components of a hi-fi system should have their ground-connection terminals connected to each other. This is usually done via the braid on the interconnecting cables. However, the ground terminal of one of the components should be connected to an external ground, such as a water pipe. The best connection at the pipe is through a clamp whose screw actually bites into the metal of the pipe. The connection should be via braid rather than ordinary wire. If a water pipe is not readily available, a radiator can be used. Another connecting point could be the center screw of the wall plate of an outlet (receptacle), assuming that the outlet is made of metal and not plastic.

A hi-fi system will work without making such a ground connection, but some hum may be present as a result. Good metal contact is needed at the site of the ground connection.

Audio Cable Noise and Hum

The audio cables joining hi-fi components may be too short or too long. If they are too short, there are several options. The easiest is to buy a cable of the right length already equipped with connectors. Another is to buy raw cable, cut it to a suitable length, and then mount connectors on each end.

Unneeded length adds to the cable confusion that can exist behind audio equipment. The longer the cable, the greater the opportunity of having the cable work as an antenna, picking up extraneous signals that are then characterized as noise and hum or bypassing treble tones because of the additional cable capacitance.

Noise-Reducing Antenna System

An *antenna system* consists of the antenna proper and its download from the signal-pickup point on the antenna to the antenna input terminals on a tuner or receiver. Both the antenna and the download are capable of signal pickup. Usually the antenna itself is mounted high enough so that it is outside the zone of noise interference. The preferable download is coaxial cable consisting of an active center conductor and a surrounding shield of metal braid. The braid should be connected to the ground terminal of the tuner or receiver. Using open wire (unshielded wire) can produce noise and hum pickup.

NOISE IN SATELLITE TV SYSTEMS

Although satellite TV systems are often considered only from the viewpoint of picture delivery, they are also systems that must supply sound. Noise interference with the picture is sometimes given the most consideration, since video noise can be much more intrusive and noticeable than that associated with audio.

3

Sound Measurements

CUSTOMARY, (ENGLISH), CGS, AND MKS SYSTEMS

There are several systems used for making sound measurements. Basically, these are the customary (practical) or English system and the metric system. The metric system is divided into two subsystems: the centimeter, gram, second (better known as the CGS) and the meter, kilogram, second (more commonly referred to as the MKS). The CGS is used when the units have small values; the MKS is used for large values. For problem solving in the metric system, either the MKS or CGS system can be used, whichever is more convenient. The units, however, cannot be intermixed and should be converted to one metric subsystem or the other prior to problem solving.

There is also a variation of the MKS system, referred to as the meter, kilogram, second, ampere (more commonly called the MKSA). Used primarily for problems in electricity and electronics, it is similar to the MKS and is interchangeable with it.

The MKS and the CGS are part of an international system of measuring units known as the Système International d'Unités, or International System of Units. The two metric systems, MKS and CGS, are used worldwide; the customary system is more common in the United States and Great Britain. The customary (formerly the English) and metric systems have conversion factors that permit interchanges of data. For example, 1 inch (customary system) is equal to 2.54 centimeters (cm) (metric system) in either the CGS or MKS. In some instances there are double conversions; for example, in the kilogram-meter (the MKS unit of work and energy) is equivalent to approximately 7.235 ft lb in the customary system.

It is sometimes difficult to determine just which system is being used. The second is used in all systems. Both metric and customary units are used for making measurements in audio. Table 3-1 supplies some useful conversion factors as an aid in working with the customary and metric systems.

TABLE 3-1 CUSTOMARY AND METRIC SYSTEM CONVERSION FACTORS

Length
  1 ft = 30.48 cm = 0.3048 m
  1 mi = 5280 ft = 1.609 km
  1 yd = 0.9144 m
  1 in. = 2.540 cm
  1 angstrom (Å) = 10^{-10} m = 10^{-8} cm
  1 fermi = 1 femtometer (fm) = 10^{-15} m = 10^{-13} cm
  1 micron (μ) = 10^{-6} m = 10^{-4} cm = 10^{4} Å
  1 light-year = 5.88×10^{12} mi = 9.461×10^{12} km
Area
  1 ft^2 = 929.0 cm^2 = 0.09290 m^2
  1 in.^{2} = 6.452 cm^2 = 645.2 mm^2
  1 barn (b) = 10^{-28} m^2 = 10^{-24} cm^2
Volume
  1 liter = 1000 cm^3 = 0.001 m^3 = 1.0576 qt = 61.03 in.^{3} = 0.0353 ft^3
  1 ft^3 = 7.481 gal = 28.32 liters
  1 m^3 = 1000 liters = 10^6 cm^3 = 1.308 yd^3 = 35.31 ft^3
Velocity
  60 mi/h = 88 ft/s = 26.82 m/s
  1mi/h = 0.4470 m/s = 1.609 km/h = 1.47 ft/s
Mass
  1 metric ton = 1000 kg = 10^6 g
Force
  1 newton (N) = 10^5 dynes = 0.2248 lb
  1 lb = 4.448 N
  1 ton = 2000 lb
  An object of mass 1 kg weighs 9.807 N = 2.205 lb
  An object of weight 1 lb has mass 453.6 g = 0.4536 kg
Pressure
  1 bar = 10^6 dyn/cm^2 = 10^5 N/m^2 = 14.50 lb/in.^{2}
  1 lb/in.^{2} = 6.89×10^4 dyn/cm^2 = 6.89×10^3 N/m^2
  1 atm = 760 torr = 760 mm Hg = 76.0 cm Hg = 14.70 lb/in.^{2} = 2116 lb/ft^2
       = 1.013×10^6 dyn/cm^2 = 1.013×10^5 N/m^2 = 1.013 bar = 1013 mbar
Work and energy
  1 joule (J) = 10^7 ergs = 0.239 cal = 0.7376 ft·lb
  1 cal = 4.184 J = 3.086 ft·lb
  1 Btu = 252 cal = 778 ft·lb = 1054 J
  1 kilowatt·hour (kWh) = 3.60×10^6 J
  1 electron volt (eV) = 1.60×10^{-19} J

Alphanumeric Identification

Letters and numbers used in acoustic formulas must sometimes require specific identification. Numbers that precede a letter or some other number are numerical coefficients. In the case of an alphanumeric arrangement such as $6ABC$, the numerical coefficient is the digit 6; this is to be multiplied by the letters. In effect, $6ABC = 6 \times A \times B \times C$. Usually the multiplying coefficient precedes the letters, but $6ABC$ can also be expressed as $ABC6$. The letters need not be supplied in the order shown but are ordinarily arranged alphabetically.

A subscript can be used for various purposes. A subscript is written to the right of another number, letter, or word. In a combination A_3, the subscript 3 does not involve any arithmetical action and is simply used to identify the letter for some specific reason. Numbers following letters need not necessarily be in subscript form, such as in R1 or C2, used in this case for identifying a specific resistor or capacitor.

A subscript is also used to identify a base. (Log_{10}) is a logarithm using 10 as a base, referred to as a *common logarithm*. A logarithm using e as a base is written (log_e) and is known as a *natural logarithm*. The value of e is about 2.718.

Unknown values in acoustic formulas are usually placed to the left of an equals sign and known values appear to the right. The square root of a number or letter can use the square root sign, as in $\sqrt{45}$, but can also be written as $(45)^{1/2}$. Both have the same meaning.

THE BEL

The *bel* is used for comparing the strengths of a pair of voltages, currents, or either acoustical or electrical powers. By itself the bel does not indicate any specific amount; it is a ratio only and is not some specific absolute value. The bel can be expressed as

$$\text{bel} = \log_{10} \frac{P_1}{P_2}$$

The ratio is logarithmic, using 10 as the base.

The Decibel

More commonly the preceding ratio is stated in terms of a tenth of a bell, or decibel (dB, formerly known as a transmission unit and used for indicating gain or loss) and is written as

$$\text{dB} = 10 \log_{10} \frac{P_1}{P_2}$$

To avoid dealing with fractions less than 1, either P_1 or P_2 can be used in the numerator. If P_2 is greater, the formula can be written as:

$$dB = 10 \log_{10} \frac{P_2}{P_1}$$

Nonlinearity

The relationship between decibel values is not linear but is logarithmic, as indicated in the formula for determining voltage, current, or power ratios. Arithmetically, 6 dB is twice as great as 3 dB, but the numbers they represent do not have a 2-to-1 relationship. A sound that is twice as strong as another has a level that is 3 dB higher. A second sound 100 times the level of another is 20 dB higher. Thus

$$10 \text{ dB} = 3.1 \times RL$$
$$20 \text{ dB} = 10 \times RL$$
$$30 \text{ dB} = 31.6 \times RL$$
$$40 \text{ dB} = 100 \times RL$$
$$50 \text{ dB} = 316 \times RL$$
$$60 \text{ dB} = 1,000 \times RL$$
$$70 \text{ dB} = 3,162 \times RL$$

where RL is the reference level (also called the zero level).

dB Power Relationships

A 1-dB sound-level increase equals 1.25 times as much audio power and is inaudible.

A 3-dB sound-level increase equals 2 times as much audio power and is just noticeably louder.

A 6-dB sound-level increase equals 4 times as much audio power and is noticeably louder.

A 9-dB sound-level increase equals 8 times as much audio power and sounds twice as loud.

dBm

The two powers being compared must be in the same units (microwatts, milliwatts, or megawatts). If either P_2 or P_1 is not known, any other power value can be selected as a reference, and while different amounts have been chosen in the past, an electrical power of 1 milliwatt (1 mW) is commonly used. In that case, the abbreviation for decibels is changed to dBm to indicate the reference power. In terms of formulas using 1 mW as the reference, the formula becomes

$$\text{dBm} = 10 \log_{10} \frac{P_1}{P_2} \qquad \text{dBm} = 10 \log \frac{0.001}{P_2}$$

or

$$\text{dBm} = 10 \log_{10} \frac{P_2}{P_1} \qquad \text{dBm} = 10 \log \frac{0.001}{P_1}$$

The equations used for the calculation of dBm make two assumptions: that the input and output impedances across which the input and output powers are measured are identical and that the output is not open-circuited. If the line is not terminated—that is, is unloaded—the dBm reference should not be used and should be replaced by dBv.

Current Measurements Using Decibels

The formula for calculating current gain or loss is expressed as

$$\text{dB} = 20 \log \frac{I_2 \sqrt{R_2}}{I_1 \sqrt{R_1}}$$

where the logarithm (log or ln) is to the base 10. Values above zero in the answer are considered positive, or a gain; those below zero are negative and represent a loss. The assumption is made that the input resistance or impedance is equal to that of the output. Under those circumstances the formula simplifies to:

$$\text{dB} = 20 \log \frac{I_1}{I_2}$$

It is important to note that input and output resistances or impedances are rarely identical in audio work. When these are not the same, use of the simplified formula will lead to errors.

Voltage Measurements Using Decibels

Voltage and current are both closely related to power, since

$$P = I^2 R \quad \text{and} \quad P = \frac{E^2}{R}$$

The voltage ratios are given by

$$\text{dB} = 20 \log \frac{E_2 \sqrt{R_1}}{E_1 \sqrt{R_2}}$$

But when the input and output resistances are equal, the formula is simplified to:

$$dB = 20 \log \frac{E_2}{E_1}$$

The simplified formulas for current and voltage ratios are applicable when input and output impedances are identical.

There are two ways for determining voltage, current, or power ratios. One method is through the use of appropriate formulas; the other is through the use of a prepared table, such as Table 3-2. The formulas and the table are mutually supportive. Using the formulas requires more work; Table 3-2 does not supply the answer for all problems. Using Table 3-2 means less work, and it can be used to verify answers obtained via formulas.

Example

The output voltage of an amplifier is 20 V; the input is 1250 mV. What is the voltage gain in dB?

Solution   Both voltages must be in the same units. 1250 mV is the same as 1.25 V. The voltage ratio is 20/1.25 = 16. There is no corresponding voltage ratio in Table 3-2. The closest ratio in Table 3-2 is 15.85, corresponding to 24 dB.

Using a formula instead of Table 3-2 gives

$$dB = 20 \log \frac{E_2}{E_1}$$

$$= 20 \log 15.85$$

$$= (20)(1.20412) = 24.08 \text{ dB}$$

Whether this answer is satisfactory or not depends on the degree of accuracy required.

NEPERS

Decibels involve logarithms having 10 as a base. The decibel is a dimensionless unit and is used for comparison purposes. However, it is also possible to use logarithms having e as a base, with that base having a value of 2.71828; such a value is identified as a neper. The ratio between a pair of currents, voltages, or power can be expressed in nepers as

$$n = \frac{1}{2} \log_e \frac{P_2}{P_1}$$

where n is the ratio in nepers, $\log_e$ is the logarithm having e as its base, and P_2 and P_1 are audio watts. P_2 and P_1 must be in the same units, watts, milliwatts, or microwatts. If not, then the units must be converted. P_1 and P_2 can be known values, or else either P_2 or P_1 can be some selected reference.

TABLE 3-2   VOLTAGE OR CURRENT RATIOS VERSUS POWER RATIOS AND DECIBELS

Voltage or current ratio	Power ratio	– dB +	Voltage or current ratio	Power ratio
1.000	1.000	0	1.0000	1.0000
0.989	0.977	0.1	1.0116	1.0233
0.977	0.955	0.2	1.0233	1.0471
0.966	0.933	0.3	1.0351	1.0715
0.955	0.912	0.4	1.0471	1.0965
0.944	0.891	0.5	1.0593	1.1220
0.933	0.871	0.6	1.0715	1.1482
0.923	0.851	0.7	1.0839	1.1749
0.912	0.832	0.8	1.0965	1.2023
0.902	0.813	0.9	1.1092	1.2303
0.891	0.794	1.0	1.1220	1.2589
0.881	0.776	1.1	1.135	1.288
0.871	0.759	1.2	1.1482	1.3183
0.861	0.741	1.3	1.161	1.349
0.851	0.724	1.4	1.175	1.380
0.841	0.708	1.5	1.189	1.413
0.832	0.692	1.6	1.202	1.445
0.822	0.676	1.7	1.216	1.479
0.813	0.661	1.8	1.230	1.514
0.803	0.646	1.9	1.245	1.549
0.749	0.631	2.0	1.2589	1.5849
0.776	0.603	2.2	1.288	1.660
0.759	0.575	2.4	1.318	1.738
0.750	0.562	2.5	1.334	1.778
0.724	0.525	2.8	1.380	1.905
0.708	0.501	3.0	1.4125	1.9953
0.692	0.479	3.2	1.445	2.089
0.676	0.457	3.4	1.479	2.188
0.668	0.447	3.5	1.4962	2.2387
0.661	0.436	3.6	1.514	2.291
0.646	0.417	3.8	1.549	2.399
0.631	0.398	4.0	1.5849	2.5119
0.596	0.355	4.5	1.6788	2.8184
0.562	0.316	5.0	1.7783	3.1623
0.531	0.282	5.5	1.8836	3.5481
0.501	0.251	6.0	1.9953	3.9811
0.473	0.224	6.5	2.113	4.467
0.447	0.200	7.0	2.239	5.012
0.422	0.178	7.5	2.371	5.623
0.398	0.159	8.0	2.512	6.310
0.376	0.141	8.5	2.661	7.079
0.355	0.126	9.0	2.818	7.943
0.335	0.112	9.5	2.985	8.913

(*continued*)

TABLE 3-2 VOLTAGE OR CURRENT RATIOS VERSUS POWER RATIOS AND DECIBELS (*continued*)

Voltage or current ratio	Power ratio	− dB +	Voltage or current ratio	Power ratio
0.316	0.100	10	3.162	10.00
0.282	0.0794	11	3.55	12.6
0.251	0.0631	12	3.98	15.9
0.224	0.0501	13	4.47	20.0
0.200	0.0398	14	5.01	25.1
0.178	0.0316	15	5.62	31.6
0.159	0.0251	16	6.31	39.8
0.141	0.0200	17	7.08	50.1
0.126	0.0159	18	7.94	63.1
0.112	0.0126	19	8.91	79.4
0.10000	0.0100	20	10.00	100.0
0.08913	0.0079	21	11.22	125.9
0.07943	0.0063	22	12.59	158.5
0.07079	0.0050	23	14.13	199.5
0.06310	0.00398	24	15.85	251.2
0.05623	0.03162	25	17.78	316.2
0.05012	0.002512	26	19.95	398.1
0.04467	0.001995	27	22.39	501.2
0.03981	0.001585	28	25.12	631.0
0.03548	0.001259	29	28.18	794.3
0.03162	0.001000	30	31.62	1000
0.02818	0.000794	31	35.48	1259
0.02512	0.000631	32	39.81	1585
0.02239	0.000501	33	44.67	1995
0.01995	0.000398	34	50.12	2512
0.01778	0.000316	35	56.23	3162
0.01585	0.000251	36	63.10	3981
0.01413	0.000199	37	70.79	5012
0.01259	0.000158	38	79.43	6310
0.01122	0.000126	39	89.13	7943
0.01000	0.000100	40	100.00	10000
0.00891	0.000079	41	112.2	12590
0.00794	0.000063	42	125.9	15850
0.00708	0.000050	43	141.3	19950
0.00631	0.000040	44	158.5	25120
0.00562	0.000032	45	177.8	31620
0.00501	0.000025	46	199.5	39810
0.00447	0.000020	47	223.9	50120
0.00398	0.000016	48	251.2	63100
0.00355	0.000013	49	281.8	79430
0.00316	0.000010	50	316.2	100000

Since the decibel is more commonly used, it is convenient to be able to convert between nepers and decimals. Some basic relationships between decibels and nepers are

$$1\ dB\ =\ 0.1\ bel \qquad 1\ bel\quad =\ 10\ dB$$

$$1\ dB\ =\ 0.1151\ neper \qquad 1\ neper\ =\ 0.8686\ bel$$

$$1\ bel\ =\ 1.151\ nepers \qquad 1\ neper\ =\ 8.686\ dB$$

A more extensive chart for conversions between decibels and nepers is supplied in Table 3-3, whereas Table 3-4 lists conversions between nepers and decibels.

TABLE 3-3 DECIBELS VERSUS NEPERS CONVERSIONS (n, nepers; dB, decibels)

dB	n	dB	n	dB	n
1	0.1151	34	3.9134	67	7.7117
2	0.2303	35	4.0285	68	7.8268
3	0.3453	36	4.1436	69	7.9419
4	0.4604	37	4.2587	70	8.0570
5	0.5755	38	4.3738	71	8.1721
6	0.6906	39	4.4889	72	8.2872
7	0.8057	40	4.6040	73	8.4023
8	0.9208	41	4.7191	74	8.5174
9	1.0359	42	4.8342	75	8.6325
10	1.1510	43	4.9493	76	8.7476
11	1.2661	44	5.0644	77	8.8627
12	1.3812	45	5.1795	78	8.9778
13	1.4963	46	5.2946	79	9.0929
14	1.6114	47	5.4097	80	9.2080
15	1.7265	48	5.5248	81	9.3231
16	1.8416	49	5.6399	82	9.4382
17	1.9567	50	5.7550	83	9.5533
18	2.0718	51	5.8701	84	9.6684
19	2.1869	52	5.9852	85	9.7835
20	2.3020	53	6.1003	86	9.8986
21	2.4171	54	6.2154	87	10.0137
22	2.5322	55	6.3305	88	10.1288
23	2.6473	56	6.4456	89	10.2439
24	2.7624	57	6.5607	90	10.3590
25	2.8775	58	6.6758	91	10.4741
26	2.9926	59	6.7909	92	10.5892
27	3.1077	60	6.9060	93	10.7043
28	3.2228	61	7.0211	94	10.8194
29	3.3379	62	7.1362	95	10.9345
30	3.4530	63	7.2513	96	11.0496
31	3.5681	64	7.3664	97	11.1647
32	3.6832	65	7.4815	98	11.2798
33	3.7983	66	7.5966	99	11.3949
				100	11.5100

TABLE 3-4 NEPERS VERSUS DECIBELS CONVERSIONS

n	dB	n	dB	n	dB
1	8.686	34	295.324	67	581.962
2	17.372	35	304.010	68	590.648
3	26.058	36	312.696	69	599.334
4	34.744	37	321.382	70	608.020
5	43.430	38	330.068	71	616.706
6	52.116	39	338.754	72	625.392
7	60.802	40	347.440	73	634.078
8	69.488	41	356.126	74	642.764
9	78.174	42	364.812	75	651.450
10	86.860	43	373.498	76	660.136
11	95.546	44	382.184	77	668.822
12	104.232	45	390.870	78	677.508
13	112.918	46	399.556	79	686.194
14	121.604	47	408.242	80	694.880
15	130.290	48	416.928	81	703.556
16	138.976	49	425.614	82	712.252
17	147.662	50	434.300	83	720.938
18	156.348	51	442.986	84	729.624
19	165.034	52	451.672	85	738.310
20	173.720	53	460.358	86	746.996
21	182.406	54	469.044	87	755.682
22	191.092	55	477.730	88	764.368
23	199.778	56	486.416	89	773.054
24	208.464	57	495.102	90	781.740
25	217.150	58	503.788	91	790.426
26	225.836	59	512.474	92	799.112
27	234.522	60	521.160	93	807.798
28	243.208	61	529.846	94	816.484
29	251.894	62	538.532	95	825.170
30	260.580	63	547.218	96	833.856
31	269.266	64	555.904	97	842.542
32	277.952	65	564.590	98	851.228
33	286.638	66	573.276	99	859.914
				100	868.600

Example

Assuming a zero reference level of 1 mW, what is the gain in nepers of an amplifier whose output is 50 mW?

Solution   The power ratio is 50 to 1. Locate the nearest number to this in Table 3-2. This is shown as 50.1. The gain in dB is 17. Now consult Table 3-3. Locate 17 dB in the left-hand column. The number of nepers corresponding to 17 dB is 1.9567.

Relative versus Absolute Measurements

Decibels represent a good example of relative measurements. Such a measurement is one that is comparative and thus simply indicates how much greater or less one unit is compared to another being used as a reference. An absolute measurement is one that has a measurement independent of a reference. A battery can be said to have a rating of 6 V. No reference is used, nor is any comparison made. The AC line voltage in a home can be 121 V. This is absolute rating, for no reference is indicated.

An absolute measurement does not require the use of a reference. A relative measurement is dependent on the selection and the use of a reference.

Technically, there is no such thing as an absolute measurement. All measurements do require two points. The difference is that with relative measurements, the reference must be supplied; with absolute measurements, the reference is implied.

REFERENCE LEVELS

In some instances when measurements are to be made, inputs and outputs are clearly specified. These must be in identical terms or else must be altered so they are. An output voltage cannot be measured against an input current, but if that input current is sent through a resistor, the result will be a voltage against which the output can be compared—that is, measured.

In some instances, when it is necessary to measure some output and no input is supplied, one can be arbitrarily selected; this value is henceforth referred to as a reference. For such measurements the reference must be understood, implied, or clearly stated.

0-dB Level

The 0-dB level is a reference for a power level based on 1 mW of power across 600 Ω. It is based on

$$W = \frac{E^2}{R}$$

where W is the power in watts, E is the voltage in volts, and R is the resistance in ohms.

The formula can be transposed to read either

$$E^2 = W \times R \quad \text{or} \quad E = \sqrt{W \times R}$$

For 1 mW = 0.001 W and R = 600 Ω,

$$E = \sqrt{0.001 \times 600} = 0.775 \text{ V}$$

0.775 V is then considered as the reference corresponding to 0 dB and is sometimes referred to as the 0-dB level. To distinguish it from other references, it is identified as dBm. Zero level is a reference power level. Thus, with 1 mW in a 600-Ω line, zero is equal to 0.775 V.

There are a number of other references, all involving dB.

dB	6 mW, 1.73 V across 500 Ω.
dBa	Used for noise measurements, with dBa representing dB adjusted.
dBf	The reference is the femtowatt, or 10^{-15} W.
dBj	The reference is 1,000 μV.
dBk	The reference is 1 kV.
dBu	The reference is 1 μV.
dBv	The reference is 1 V.
dBW	The reference is 1 W.
dBx	Used for cross talk measurements.
dBrap	For dB above the reference acoustical power.
dBrnc	Used for cross talk measurements. A dBx reference can also be used.
dBVg	Decibels of voltage gain.
dBmV	1,000 μV (or 1 mV) can be used in connection with the decibel supplying these figures as equivalent to 0 dB.

Typically, the reference level is 0 dB = 0.001 W, 0.006 W, or 0.0125 W.

Volume Units

A volume unit (VU) indicates a change of 1 dB in volume and is used in conjunction with complex waveforms. Amplitude changes in such waveforms can be measured only in VU.

Volume units can be measured by ballistic-type VU meter. Other types, such as LED, LCD, and fluorescent bar displays measure dBs, not VUs.

MEASUREMENTS OF SOUND PRESSURE

The Bar

The bar (an abbreviation of barometric) is a CGS unit of atmospheric pressure and is equal to 10^6 dyn/cm². Its equivalent in the English system is 14.7 lb/in.² at sea level. Sea level is selected as the reference, since atmospheric pressure decreases with altitude. A millibar (mbar) is a thousandth of a bar and is equivalent to 10^3 dyn/cm².

Sound pressure, either vocal or music, at the input of a microphone is simply a variation of the existing atmospheric pressure. The sound pressure put on a microphone is much smaller than atmospheric pressure and is measured in terms of a millionth of an atmosphere or in microbars. This is still substantially greater than the threshold of hearing, the point at which sound becomes perceptible, which is equivalent to 0.0002 μbar.

Since the bar is a unit of pressure, it is measured in terms of the area over which it exists. The formula is

$$P = \frac{F}{A}$$

where P is the pressure, F is the force, and A is the area over which the force is applied perpendicular to the area. The units used in this equation can be customary or metric. The equivalents of 1 atm are given in Table 3-5.

Variations in Pressure Measurements

There are a number of ways in which pressure can be stated, such as N/m^2, dyn/cm^2, lb/ft^2, lb/in.^{2}, and so on, where N is newtons, m is meters, dyn is dynes, cm is centimeters, lb is pounds, ft is feet and in. is inches. Pressure is sometimes indicated in atmospheres (atm), where 1 atm is representative atmospheric pressure at sea level. An atmosphere is 14.7 lb/in.^{2}, 1.013×10^5 N/m, or 1.013×10^6 dyn/cm^2. More often, in audio, the common unit of pressure (also used in meteorology) is the bar, representing a pressure of 10 dyn/cm^2. Table 3-6 lists various units of pressure for 1 atm.

Sound pressure is obtained from any moving body, such as the strings of a violin or the membrane of a drum, or any method which disturbs the atmosphere. The variation in sound pressure can be at an audio rate. Sound pressure is also measured in pascals.

Pascal

The pascal (Pa) is another unit used for measuring sound pressure and is equivalent to 10 microbars, or 10 dyn/cm^2, and is approximately equal to 94 dB sound pressure level. The pascal is also equal to 1 Newton per square meter (N/m^2). Like

TABLE 3-5 EQUIVALENTS OF 1 ATMOSPHERE

To convert	Into	Multiply atmospheres by	Conversely multiply by
Atmospheres	bars	1.0133	0.9869
Atmospheres	mm of mercury at 0°C	760	1.316×10^{-3}
Atmospheres	ft of water at 4°C	33.9	2.950×10^{-2}
Atmospheres	in. of mercury at 0°C	29.92	3.342×10^{-2}
Atmospheres	kg per m^2	1.033×10^4	9.678×10^{-5}
Atmospheres	N per m^2	1.0133×10^5	0.9869×10^{-5}
Atmospheres	lb per in.^{2}	14.70	6.804×10^{-2}

TABLE 3-6   UNITS OF PRESSURE FOR 1
ATMOSPHERE

$$1 \text{ atm} = 14.7 \text{ lb/in.}^2$$
$$= 1.013 \times 10^5 \text{ N/m}^2$$
$$= 1.013 \times 10^6 \text{ dyn/cm}^2$$
$$= 1.013 \text{ bar}$$
$$= 76 \text{ cm Hg}$$
$$= 760 \text{ mm Hg}$$
$$= 760 \text{ torr}$$
$$= 34 \text{ ft of water}$$

the pascal, the microbar is used for measuring sound pressure and is one millionth (10^{-6}) of normal atmospheric pressure.

The Dyn

There are various ways of defining air pressure, including the microbar, SPL, the dyn, and the pascal. All of these are units of force. The dyn, a unit in the CGS system, is defined as the force that will produce a velocity of 1 cm/sec when acting on a mass of 1 gram.

The dyn/cm² is the equivalent of 1 μbar, or 74 dB sound pressure level. Ten dyn/cm² is the equivalent of 10 μbars, or 94 dB sound pressure level (SPL).

The Newton

Just as the dyn is a unit of force in the CGS system, the newton (N) is a unit of force in the MKS system. In this system the mass is the kilogram (kg); the gram (g) is the unit of mass in the CGS. The unit in which the acceleration is expressed is the meter per second (m/s). It is possible to convert from dyns to newtons because 1 N = 10 dyn.

Either the CGS or MKS systems can be used for the determination of units of force. The MKS is more convenient to use when the measurements are relatively large; the centimeter is more useful in problems in which the measurements are small.

Both MKS and CGS are metric. For the customary, English or practical system, the unit of force is the pound.

MICROPHONE MEASUREMENTS

Table 3-7 lists the units used in making measurements of microphones, where mV is millivolts (10^{-3} V, or 0.001 V); Pa is pascals, with 1 Pa = 10 μbar or 10 dyn/cm², approximately equal to 94 dB SPL; μb is microbars; V is volts; dB is decibels;

TABLE 3-7 MICROPHONE MEASUREMENTS

mV/Pa	Millivolts per pascal is equivalent to ($\triangleq$) mV/N/m^2 = output voltage per rated sound pressure of 1 Pa (1 Pa $\triangleq$ 10 μb)
mV/μb	Millivolts per microbar $\triangleq$ 10 mV/Pa = output voltage per rated sound pressure of 1 microbar (1 μb $\triangleq$ 0.1 Pa)
dBV	Output level related to 1 V
Hz	Hertz = 1 cycle/sec = unit of frequency
ohm	Ω = unit of impedance
dB SPL	dB sound pressure level (SPL), related to the SPL of 20 μPa or 2 $\times$ 10^{-4} μb and measure of equivalent noise level
μV/μt	Microvolts per microtesla = unit of hum sensitivity. 5 μt = 50 mG = related hum field
dBm	Output level related to 0.775 V at 600 Ω = 1 mW at 600 Ω
dyn	The dyn per square centimeter (dyn/cm^2) = 1 microbar (1 μbar) or 74 db SPL. 10 dyn/cm^2 = 10 μB or 94 dB SPL.
0 dBu	Defined as 0.775 V regardless of load impedance. Subtract 2.2 from the dBu figure to convert to dBu referenced to 1 V. When the load impedance is 600 Ω, this particular dBu is also known as dB/m.

Hz is equivalent to cycles/second; SPL is the sound pressure level; and μT is microteslas.

Need for a Reference in Measurements

A measurement is a comparison between a pair of values, one of which is used as a reference. In this sense it is also a ratio, a measurement between two points. A distance of 1 mi is a measurement between two points, one of which is arbitrarily selected as zero. In making voltage measurements, it is essential to have a starting point, or reference, and this can be any arbitrarily selected value. It is important, however, to make clear just what the reference point is.

In the case of units involved with a pre- or power amplifier, the input is an audio AC voltage and is the reference. The output is also an AC audio voltage, and the ratio of the output to the input is the gain of the amplifier. Unlike amplifiers, though, for microphones the input is not a voltage but is a force, with that force expressed in microbars, dynes, or newtons. These are all in the metric system, since the English system is not used for such measurements.

Since the output of the microphone is in volts, the input can also be stated in volts, but the input pressure and the voltage that results must be considered.

It is possible for a microphone to have an output that is less than the selected reference. This could be the case with low-impedance microphones, which ordinarily produce an output that is smaller than its reference. If the output-input comparison is in terms of voltage, decibels are used for establishing the ratio of output to input. Since the output is smaller than the input, the output is in negative decibels. This does not indicate subtraction but simply means that the price paid for the use of a microphone as a transducer is a drop in voltage. The pressure that is used at the input to the microphone is usually in microbars or dyn/cm^2.

The ratio of output to input is a measure of the sensitivity of a microphone. The sensitivity could be listed in a spec sheet as

$$-74 \text{ dB re 1 V/}\mu\text{bar}$$

This means that the output of the microphone is -74 dB compared to a reference of 1 V/μbar. Each microbar of pressure on the diaphragm results in an electrical pressure of 1 V.

Alternatively, the sensitivity of a different microphone could be shown as

$$-58 \text{ dB re 1 mW/10 dyn/cm}^2$$

The reference in this example is 1 mW. The output is 10 dyn/cm^2. Ten dynes is the equivalent of 10 μbar. The input is 1 mW per 10 dyn force on each square centimeter of the microphone's diaphragm.

Since 10 dyn/cm^2 is the same as 10 μbar, the preceding statement can also be written as

$$-58 \text{ dB re 1 mW/10 }\mu\text{bar}$$

All the microphones just listed have the same output for an equivalent input. Consequently, they all have the same sensitivity and will produce the same output when the input is identical under all conditions. Note, though, that the first microphone, using a larger negative reference, would appear to supply a larger ratio between input and output. But the output is the same in all three examples, although expressed differently.

The output level of a microphone has no value unless the specific frequency being used and the microphone impedance are also supplied. To determine the sensitivity of a microphone in comparison with others, all input conditions must be the same.

LOUDNESS

Loudness is the extent to which the human ear recognizes an audio stimulus. Basically, loudness depends on the sound pressure level of the source. Loudness is dependent on frequency; for human hearing, sensitivity is more responsive to middle-frequency tones but is less sensitive to sounds having lower and higher frequencies.

The loudness of a sound is subject to a large variety of influences in addition to frequency. The intensity of the sound, its timbre, the time duration of a sound, plus the hearing ability of the individual listener all affect the loudness of a sound.

SOUND PRESSURE LEVEL

Sound pressure level (SPL) is an alternative to the microbar for use as a unit of sound pressure measurement. The threshold of hearing is 0 dB SPL, corresponding to 0.0002 μbar or 0.00002 N/m². SPL measurements are made in decibels, but it can also be expressed as intensity in terms of watts per square meter.

Sound is produced by very small changes in SPL. Barely audible sound is caused by an increase in SPL of about 10%. A sound level of 130 dB requires an increase of SPL of about 0.1% to put the sound at the threshold of pain.

SOUND INTENSITY LEVEL

The human ear responds to an extremely wide range of sound intensities but is fairly insensitive to changes in those intensities. Sound intensity is defined as the power in watts through 1 cm² with the unit of sound intensity equal to 10 W/cm².

Sound intensity level can range from the threshold of hearing to the threshold of pain, with the intensity measured in W/m². Sound intensity can be defined as

$$SI = 10 \log_{10} \frac{I}{I_0}$$

where SI is the sound intensity in decibels and I is the intensity at the threshold of hearing; I_0 is the reference.

When measured at a frequency of 1,000 Hz,

$$I_0 = 10^{-12} \text{ W/m}^2$$

where I is the intensity of the sound at any level above the threshold of sound—that is, at dB = 0, W is the power in watts, and m is the area in square meters. I_0 is the reference.

Loudness, as perceived by the human ear, is based on the amount of radiated acoustical energy and more explicit is equal to its intensity. Intensity is defined as the ratio of acoustic power to the area. Expressed as a formula,

$$I = \frac{P}{m^2}$$

where I is the intensity, P is the power in watts, and m^2 is the area in square meters. Since I is proportional to P, the greater the amount of radiated acoustical power, the greater is the perceived intensity. I is also inversely proportional to m^2, thus indicating that as the area through which the sound field moves is increased, the sound intensity decreases. If the distance from the sound source is doubled, the sound intensity is decreased by a factor of 4; if distance is tripled, intensity decreases by a factor of 9. This indicates that the sound intensity is inversely proportional to the square of the distance. Expressed as a formula,

$$I = \frac{1}{d^2}$$

where I is the perceived intensity and d is the distance.

Intensity in Terms of Musical Notation

Although the intensity of sound can be expressed mathematically in terms of watts per square meter, it is also indicated by letters as shown in Table 3-8.

TABLE 3-8   INTENSITY VERSUS LOUDNESS

Intensity (W/m^2)		Loudness
1		Threshold of feeling
10^{-3}	*fff*	Extremely loud
10^{-4}	*ff*	Very loud
10^{-5}	*f*	Loud
10^{-6}	*mf*	Moderately loud
10^{-7}	*p*	Soft
10^{-8}	*pp*	Very soft
10^{-9}	*ppp*	Extremely soft
10^{-10}		Threshold of hearing

The Sone

The sone is a unit for measuring loudness—that is, the hearing characteristics of the human ear. In terms of a formula,

$$1 \text{ sone} = 10 \text{ } \mu W/cm^2$$

The loudness of a tone at a frequency of 1 kHz that is 40 dB above the threshold of hearing is 1 sone. This is used as the reference. A tone that is twice as loud is 2 sones, assuming both tones are measured at the same frequency. A thousandth of a sone is referred to as a millisone:

$$1 \text{ millisone} = 10^{-3} \text{ sone}$$

A millisone is sometimes referred to as a *loudness unit*.

A volume control on a receiver is not the same as a loudness control. The purpose of the volume control is to adjust the overall sound level over the entire audio spectrum. Further, the volume control is not frequency-selective. The purpose of the loudness control is to compensate for our hearing. Our hearing is relatively insensitive to bass and treble tones when the overall volume is low. The loudness control boosts these tones at low-volume settings.

$$SI = 10 \log \frac{I}{I_0} = 10 \log \frac{10^{-12} \text{ W/m}^2}{10^{-12} \text{ W/m}^2}$$

$$= 10(\log 1) = 10(0) = 0 \text{ dB}$$

Figure 3-1 shows the various situations that supply different levels of sound intensity (SI), starting with 0 dB at the bottom, representing the threshold of hearing, and 130 dB at the top, with sound so intense that it produces listening pain. At the threshold of feeling, the sound intensity is in the region of 120 dB to

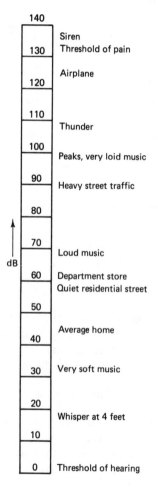

Figure 3-1 Sound range in decibels.

130 dB. This threshold can be calculated by

$$SI = 10 \log \frac{I}{I_0} = 10 \log \frac{1 \text{ W/m}^2}{10^{-12} \text{ W/m}^2}$$

$$= 10 \log \frac{1 \times 10^{12}}{10^{-12} \times 10^{12}} = 10 \log \frac{10^{12}}{1}$$

$$= 10(\log 10^{12}) = 10(12) = 120 \text{ dB}$$

Nearly all sounds come within the 0-dB to 120-dB range, but some are even more intense and are capable of an intensity of 130 dB to 140 dB. While the sound pressure is measured in dynes per square centimeter, the audio power at the entrance to the ear will be in watts per square centimeter.

The aural range for the human ear extends from 1 W/m^2, the point at which the perceived sound is painful, to $10°$, with the sound barely heard. Exponentially, the digit 1 can be expressed as $10°$. Similarly, -10^{12} is equivalent to 1 followed by 12 zeros, preceded by a minus sign. This is a million million and supplies an indication of the tremendous range of sound perception by the human ear.

Sound intensity level (SI) appears in a formula as

$$SI = 10 \log_{10} \frac{I_0}{I_{in}}$$

where I_{in} is the intensity at the threshold of hearing, I_0 is the comparison level, and SI is the sound intensity level in decibels.

The response of the ear isn't the same at all audio frequencies but peaks in the audio range of 1 to 2 kHz for most people.

The Phon

The phon is a unit of loudness and as such can be used to replace or supplement alphabetic units used in marking music, as indicated in Table 3-9. As a unit, the phon is equivalent to the smallest change in sound intensity apparent to the ear.

The scale of phons starts at zero, representing the faintest audible sound. This scale corresponds to the decibel scale of sound intensity. The number of phons of a given sound corresponds to the number of decibels of a pure tone

TABLE 3-9 MUSICAL UNITS OF LOUDNESS
VERSUS PHONS

ppp	20 phons	*f*	75 phons
pp	40 phons	*ff*	85 phons
p	55 phons	*fff*	95 phons
mf	65 phons		

having a reference frequency of 1,000 Hz. Due to the nonlinear hearing response of the human ear, doubling the intensity of a sound does not double the sound level that is heard. A true loudness scale is one that doubles the sound sensation heard when the intensity is doubled. This is a characteristic of the phon scale.

Sones can be converted into phons through the use of this equation:

$$S = 2^{(P-40)/10}$$

where S is the loudness is sones and P is the loudness level in phons. The loudness level in phons is the equivalent of the sound pressure in decibels relative to 2×10 μbar at 1 kHz. Using this as a reference, 40 phons = 1 sone. Table 3-10 is useful for making conversions between sones and phons.

TABLE 3-10 CONVERSIONS BETWEEN PHONS AND SONES

Phons	0	+1	+2	+3	+4	+5	+6	+7	+8	+9
	Sones									
20	0.25	0.27	0.29	0.31	0.33	0.35	0.38	0.41	0.44	0.47
30	0.50	0.54	0.57	0.62	0.66	0.71	0.76	0.81	0.87	0.93
40	1.0	1.07	1.15	1.23	1.32	1.41	1.52	1.62	1.74	1.87
50	2.0	2.14	2.30	2.46	2.64	2.83	3.03	3.25	3.48	3.73
60	4.0	4.29	4.59	4.92	5.28	5.66	6.06	6.50	6.96	7.46
70	8.0	8.60	9.20	9.80	10.6	11.3	12.1	13.0	13.9	14.9
80	16.0	17.1	18.4	19.7	21.1	22.6	24.3	26.0	27.9	29.9
90	32.0	34.3	36.8	39.4	42.2	45.3	48.5	52.0	55.7	59.7
100	64.0	68.6	73.5	78.8	84.4	90.5	97.0	104	111	119
110	128	137	147	158	169	181	194	208	223	239
120	256	274	294	315	338	362	388	416	446	478

Phons are shown in this table by the first column at the left and the first row across the top. Thus, 40 phons is equivalent to 1.0 sone. To obtain the equivalent of 41 phons, add 40 phons to +1 phon across the top, giving 1.07 sones.

UNITS OF FREQUENCY

At one time frequency was specified in terms of cycles per second (cps). This was subsequently replaced by the hertz, an equivalent in the International System of Units (SI). Thus, a frequency of 100 cps is now written as 100 Hz. Table 3-11 lists various ways of expressing frequency.

1 Hz is a complete cycle of 360°. The time element is always 1 second. Thus, if a single cycle is completed in 1 second the frequency is 1 Hz. A frequency of 1 kHz means that 1,000 cycles have been completed in 1 second.

TABLE 3-11 CONVERSIONS OF FREQUENCY UNITS

1 Hz	= 1/1,000 kHz = 10^{-3} kHz = 1/1,000,000 MHz = 10^{-6} MHz
1 kHz	= 1,000 Hz = 10^3 Hz = 10^{-3} MHz = 1/1,000 MHz
1 MHz	= 1,000,000 Hz = 10^6 Hz = 10^3 kHz = 1,000 kHz

PERIODIC MOTION

Any motion that repeats itself regularly is identified as periodic. The movement of a pendulum is periodic, as is the movement of the arm of a metronome. The motion of the sweep second hand of a watch is periodic. The voltage waveform supplied by an AC power line to a receptacle in a home is periodic.

A wave can be periodic or nonperiodic. Music is periodic, whereas noise is nonperiodic. Periodic waveforms have a number of specific characteristics: period, or t, the time for the completion of a single wave; frequency, the number of complete waves per unit time; amplitude, a measurable amount of maximum strength; wavelength, the distance between the start of a single cycle and its completion; and velocity.

One of the parameters is the frequency, or the number of complete cycles of the motion per unit of time. Time can be in hours, days, or minutes, but the second is often used as the time unit.

The number of complete cycles is inversely proportional to the selected time unit and can be expressed as

$$f = \frac{1}{t}$$

where t is the time in seconds and f is the frequency, or the number of complete cycles per second.

Since frequency and time have an inverse relationship, the higher the frequency, the smaller the amount of time for the completion of a single complete cycle of that frequency.

Still, another parameter is the extent of the motion. A pendulum may move 2 or 3 cm about its center, or resting point; the displacement of the violin string is so small that it is impossible to observe directly.

The time duration of a single waveform of a periodic wave is inversely proportion to its frequency in hertz. Thus,

$$t = \frac{1}{f}$$

where t is the time in seconds and f is the frequency in hertz. When the frequency is to be in kilohertz, divide f by 1,000; when the frequency is to be in millihertz, divide f by 1,000,000. To convert seconds to milliseconds, multiply seconds by

1,000; to convert seconds to microseconds, multiply seconds by 1,000,000. The relationship between the frequency of a periodic waveform and the time for the completion of a single cycle of a periodic waveform is supplied in Table 3-12.

TABLE 3-12 TIME VERSUS FREQUENCY OF A PERIODIC WAVEFORM

Frequency	Microseconds	Milliseconds	Seconds
1.0 Hz	1,000,000	1,000	1.0
10	100,000	100	0.10
50	20,000	20	0.02
100	10,000	10	0.01
200	5,000	5	0.005
500	2,000	2	0.002
1 kHz	1,000	1.0	0.001
2	500	0.5	0.0005
5	200	0.2	0.0002
10	100	0.1	0.0001
20	50	0.05	0.00005
50	20	0.02	0.00002
100	10	0.01	0.00001
200	5	0.005	0.000005
500	2	0.002	0.000002
1 MHz	1.0	0.001	0.000001
2	0.5	0.0005	0.0000005
5	0.2	0.0002	0.0000002
10	0.10	0.0001	0.0000001
20	0.05	0.00005	0.00000005

Thus, the time period for a single cycle of a 500-Hz waveform is 1/500 = 0.002 second. This is equivalent to 0.002 × 1,000, or 2 milliseconds. It is also equivalent to 0.002 × 1,000,000 µs, or 2,000 microseconds.

A musical waveform often appears to be complex and not likely to be designated as periodic. The waveform can be synthesized by Fourier analysis and shown to consist of a fundamental wave and a number of harmonics, all of which are sine waves and hence periodic.

RESONANCE

Resonance is usually a large mechanical vibration caused by a small periodic stimulus. The amplitude of the vibration will depend on how close the frequency of the stimulus is to the natural vibration time of the object subjected to the transfer of energy. A tuning fork, struck and suspended in space, results in a very small sound level, due to its very small surface area. There will be a considerable increase in sound volume if the fork is touched to a large surface, such as a table. The larger sound does not mean an increase in energy. Suspended in space the tuning fork will continue to vibrate for some time after the vibration of the large surface ceases. The increase in sound is caused by the fact that the surface is in contact with a much greater amount of air molecules. The greatest amount of transferred sound will exist when the natural resonant frequency of the plane surface is the same as that of the tuning fork.

Determination of Resonance Frequencies

A room's resonance frequency can be determined by

$$f = \frac{C}{L}$$

where f is the frequency in hertz, C is the speed of sound in feet per second, and L is the wavelength in feet. C is generally used to represent the speed of a wave, such as that of sound or light. At room temperature the speed of sound is about 1,128 ft/s.

Assume a room whose dimensions are 20 ft $\times$ 12 ft. For the 20-ft dimension,

$$f = \frac{1,128}{20} = 56.4 \text{ Hz}$$

Consequently, there will be a resonance at this frequency. This is the fundamental, so there will also be resonance frequencies at the harmonics—that is, at 56.4 $\times$ 2 = 112.8 Hz, 56.4 $\times$ 3 = 169.2 Hz, and so on. For the 12-ft dimension the resonance frequencies will be 1,128/12 = 94 Hz, 94 $\times$ 2 = 188 Hz, 94 $\times$ 3 = 282 Hz, and so on.

If both wall dimensions have a resonance frequency in common, there will be a strong resonance.

Tuner or Receiver Input Sensitivity

The sensitivity of a tuner or receiver can be measured as a ratio between the output and input voltages, or the output versus input (signal) power. Specs have been changed to favor power ratings. When the sensitivity figure is specified in decibels, it is understood that voltage is used as the reference measurement. To call attention to the fact that power is the reference, the abbreviation used is dBf, where f indicates the femtowatt, equivalent to 10^{-15} W. Table 3-13 lists tuner or receiver input sensitivity in terms of dBf.

If the receiver is an AM/FM type, sensitivity figures for both sections are different and so are listed separately. This spec indicates the amount of input signal voltage required for useful speaker output.

If the voltage involved in the comparison is negative, a minus sign is used preceding dBmV, as shown in Table 3-14. For all values above 1,000 μV, dBmV are written as positive numbers, and with these the plus polarity symbol is not

TABLE 3-13 TUNER OR RECEIVER INPUT SENSITIVITY (μV) VERSUS dBf

μV	dBf	μV	dBf	μV	dBf
1.5	8.71	2.6	13.478	3.7	16.548
1.6	9.28	2.7	13.804	3.8	16.776
1.7	9.8	2.8	14.134	3.9	17.012
1.8	10.3	2.9	14.436	4.0	17.2
1.9	10.77	3.0	14.74	5.0	19.17
2.0	11.198	3.1	15.01	10.0	25.19
2.1	11.6	3.2	15.3	30.0	34.74
2.2	12.04	3.3	15.7	40.0	37.23
2.3	12.424	3.4	15.8	50.0	39.17
2.4	12.8	3.5	16.07	100.0	45.19
2.5	13.15	3.6	16.3	1000.0	65.19

used. This table is convenient, but for values not covered, the conversion of sensitivity in terms of microvolts per meter can be used:

$$dBf = 20 \log\left(\frac{\mu V/m}{0.55}\right)$$

where log is log to the base 10 and μV is microvolts.

OPEN-REEL DECKS

At one time open-reel decks were fairly popular in the home, but these units are now mostly used in recording studios and by professional musicians who require multitrack simul-sync and the ease with which open-reel tape can be edited, something difficult or impossible with cassette tape. Another advantage is the greater amount of headroom available for open reel.

CASSETTE TAPE CLASSIFICATIONS

There are four classifications of cassette tape, with each type identified by a Roman numeral.

Type I. Type I tape uses ferric oxide particles, requires lower bias than the other types, and requires 120 μs equalization.

Type II. Type II tape uses chromium dioxide tape or any one of its equivalent formulations, such as cobalt-doped ferric oxide. Requires 70 μs equalization and higher bias than that used for Type I.

TABLE 3-14 dBmV VERSUS μV

dBmV	μV	dBmV	μV	dBmV	μV
-40	10	0	1,000	40	100,000
-39	11	1	1,100	41	110,000
-38	13	2	1,300	42	130,000
-37	14	3	1,400	43	140,000
-36	16	4	1,600	44	160,000
-35	18	5	1,800	45	180,000
-34	20	6	2,000	46	200,000
-33	22	7	2,200	47	220,000
-32	25	8	2,500	48	250,000
-31	28	9	2,800	49	280,000
-30	32	10	3,200	50	320,000
-29	36	11	3,600	51	360,000
-28	40	12	4,000	52	400,000
-27	45	13	4,500	53	450,000
-26	50	14	5,000	54	500,000
-25	56	15	5,600	55	560,000
-24	63	16	6,300	56	630,000
-23	70	17	7,000	57	700,000
-22	80	18	8,000	58	800,000
-21	90	19	9,000	59	900,000
-20	100	20	10,000	60	1.0 V
-19	110	21	11,000	61	1.1
-18	130	22	13,000	62	1.3
-17	140	23	14,000	63	1.4
-16	160	24	16,000	64	1.6
-15	180	25	18,100	65	1.8
-14	200	26	20,000	66	2.0
-13	220	27	22,100	67	2.2
-12	250	28	25,000	68	2.5
-11	280	29	28,000	69	2.8
-10	320	30	32,000	70	3.2
-9	360	31	36,100	71	3.6
-8	400	32	40,000	72	4.0
-7	450	33	45,000	73	4.5
-6	500	34	50,000	74	5.0
-5	560	35	56,000	75	5.6
-4	630	36	63,000	76	6.3
-3	700	37	70,000	77	7.0
-2	800	38	80,000	78	8.0
-1	900	39	90,100	79	9.0
-0	1,000	40	100,000	80	10.0

Type III.   Type III tape is double-layered tape in which a ferric oxide layer has a second layer of chromium dioxide particles positioned over the ferric oxide. This tape is seldom used.

Type IV.   Type IV tape is pure metal tape. It requires a higher bias than any of the other types. Equalization is 70 μsec.

Hum Sensitivity

Hum sensitivity refers to the sensitivity of a microphone to induced hum voltages. If the microphone is phantom powered from some component that depends on line power, there is the possibility of hum voltages using this for hum-signal transfer. More commonly, the hum voltage is induced by stray magnetic fields. Hum sensitivity is expressed in terms of microvolts per microtesla:

microvolts per microtesla − unit of hum sensitivity = related hum field μV/μT

$$5 \ \mu T \geqq 50 \ mG$$

The prefix μ stands for micro and is equal to 10^{-6}; T is the tesla, the unit of magnetic flux density in the MKS system, which is equivalent to 1 Wb/m² (the weber (Wb) is the same as 10^8 maxwells, or 100,000,000 lines of magnetic flux); and G is the gauss, a unit of magnetic induction. The formula for calculating the unit of hum sensitivity involves induced hum voltage only and does not include any hum voltage carried by a power line.

$$- \ 10 \ mV/Pa \ = \ \frac{\text{output voltage per rated sound pressure of 1 microbar}}{(1 \ \mu bar \geqq 0.1 \ Pa) \ mV/\mu bar}$$

Inverse Square Law

The *inverse square law* describes how microphone output (or loudspeaker output) decreases as the listener moves away from the microphone or loudspeaker. The law states that the sound pressure will drop 6 dB each time the distance is doubled. Strictly speaking, this law applies only to a situation where no reverberation is present, as in a large field out-of-doors. However, the law still holds true for any normal microphone working distance and is also true for distances from a loudspeaker up to about 10 ft. The precise distance is dependent on room acoustics, increasing for "dead" rooms and decreasing for highly reverberant rooms.

$$mV/N/m^2 \ = \ \text{millivolts per pascal}$$

$$= \ \text{output voltage per rated sound pressure of 1 Pa}$$

$$(1 \ Pa \geqq \mu bar) \ mV/Pa$$

where mV is millivolts (10^{-3} V), N is newtons, m is meters, Pa is pascals, and μbar is microbars.

THE MEL

The *mel* is the subjective unit of musical pitch. One thousand mels is the subjective perception of a 1-kHz tone having an SPL of 40 dB above a reference level of 0 dB, or the threshold of hearing.

WAVELENGTH VERSUS FREQUENCY

A sound wave having a constant velocity will travel a distance of one wavelength in an interval of one period. This can be expressed as

$$\lambda = c \times t$$

where λ is the wavelength, c is the speed of sound, and t is the period.

There are various ways of measuring a wave, as illustrated in Figure 3-2. These include the wavelength, the phase and the period. The sine wave shown here is a basic waveform and cannot be resolved into a combination of other wave types.

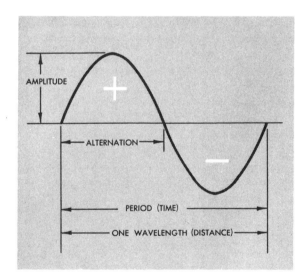

Figure 3-2 Wavelength can be measured in terms of time or distance.

The frequency of a sound wave and its wavelength have an inverse relationship expressed by:

$$f = \frac{1}{\lambda} \quad \text{or} \quad \lambda = \frac{V}{f}$$

where λ is the wavelength, f is in hertz, and V is the velocity in either feet per second or meters per second. If velocity is in feet per second, the wavelength is in feet; if it is meters per second, the wavelength is in meters. Table 3-15 supplies a comparison of wavelength versus frequency, where the wavelength is in feet or inches and the frequency range is from 20 Hz to 20 kHz.

TABLE 3-15 WAVELENGTHS OF SOUND

		(1130 ft/s, in air, at 20°C; 32° F)			
Frequency (Hz)	Wavelength (ft)	Frequency (Hz)	Wavelength (ft)	Frequency (Hz)	Wavelength (ft)
20	56.50	140	8.07	380	2.97
25	45.20	150	7.53	400	2.83
30	37.67	160	7.06	420	2.69
35	32.29	170	6.65	440	2.57
40	28.25	180	6.28	460	2.46
45	25.11	190	5.95	480	2.35
50	22.60	200	5.65	500	2.26
55	20.55	210	5.38	525	2.15
60	18.83	220	5.14	550	2.05
65	17.38	230	4.91	575	1.97
70	16.14	240	4.71	600	1.88
75	15.07	250	4.52	650	1.74
80	14.13	260	4.35	700	1.61
85	13.29	270	4.19	750	1.51
90	12.56	280	4.04	800	1.41
95	11.89	290	3.90	850	1.33
100	11.30	300	3.77	900	1.26
110	10.27	320	3.53	950	1.19
120	9.42	340	3.32	975	1.16
130	8.69	360	3.14	990	1.14
Frequency (Hz)	Wavelength (in.)	Frequency (Hz)	Wavelength (in.)	Frequency (Hz)	Wavelength (in.)
1,000	13.56	9,000	1.51	16,000	0.85
2,000	6.78	10,000	1.36	17,000	0.80
3,000	4.52	11,000	1.23	18,000	0.75
4,000	3.39	12,000	1.13	19,000	0.71
5,000	2.71	13,000	1.04	20,000	0.68
6,000	2.26	14,000	0.97		
7,000	1.94	15,000	0.90		
8,000	1.70				

Wavelength versus Frequency in Musical Instruments

As indicated by the formula shown previously, wavelength and frequency have an inverse relationship; consequently, the shorter the wavelength, the higher the frequency. This relationship has a practical application in the design of a musical instrument, because the size of the instrument determines the longest wavelength it can generate. Since a flute is quite small, it produces high-frequency tones, that is, tones that have a short wavelength. A tuba produces long-wavelength sounds.

VELOCITY OF SOUND

Sound travels through the air with a speed of about 766 m/h or 1,117 ft/s at sea level. It is affected by temperature, as indicated in the following formulas. The first of these uses the customary system of measurement; the second involves the metric system.

$$V = 49 \sqrt{459.4} + F \text{ ft/s}$$

$$V = 20.06 \sqrt{273} + C \text{ m/s}$$

where V is the velocity in feet per second or in meters per second, F is the temperature in degrees Fahrenheit; and C is the temperature in degrees Celsius.

The velocity of sound is fairly independent of its frequency, is affected slightly by humidity, but is substantially dependent on the material on which it is incident and through which it moves. Table 3-16 supplies data relating the velocity of sound through the air and the temperature in °F and °C. This Table indicates that the velocity of sound rises with increases in temperature.

TABLE 3-16 VELOCITY OF SOUND IN AIR

°F	Speed (ft/s)	°C	Speed (m/s)
32	1087	0	331.32
50	1107	10	337.42
59	1117	15	340.47
68	1127	20	343.51
89	1147	30	349.61

Time Lag versus Velocity

Music listeners in the far forward rows of a music hall hear direct sound earlier than listeners in the balcony, with the sound separation referred to as *time lag*. Time lag includes the time needed for the direct sound to reach a reflecting surface plus the time needed for the reflected sound to reach the ears of the listener.

Velocity versus Speed of Sound Waves

Although velocity and speed are often considered synonymous when used as descriptive terms in discussions of sound, they are different. Velocity means the oscillating velocity of a given particle of the medium with reference to the medium as a whole. For speed, the unit of measurement is the centimeter per second. The speed of an object is an indication of how fast it moves. In terms of a formula,

$$\text{Speed} = \frac{\text{distance traveled}}{\text{elapsed time}}$$

Velocity can be expressed as:

$$v_{\text{av}} = \frac{s_f - s_i}{t_f - t_i}$$

where v_{av} is the average velocity, s_i is the initial position at time t_i, and s_f is the final position at time t_f. Thus, an average velocity is the displacement of $s_f - s_i$ taking place in the time interval $t_f - t_i$.

If $s_i = 0$ when $t_i = 0$ and s_f is the displacement s when $t_f = t$, then

$$v_{\text{av}} = \frac{s}{t}$$

If the distance traveled and the displacement are the same, the average velocity and the average speed will be numerically identical. But if the distance traveled is larger than the displacement, the average speed will be numerically larger than the average velocity.

Calculations of the velocity of sound are based on the premise that the sound field is uniform. However, the differences are fairly small and so the formula for sound velocity considers the field to be uniform overall.

The velocity of sound in air is dependent on temperature, atmospheric pressure (ATM), and the amount of moisture in the air, although this latter factor is commonly ignored.

Effect of Frequency on the Velocity of Sound

Frequencies in the sound spectrum travel at the same speed through air. If that were not the case, then the various instruments of an orchestra would be heard at different time intervals. The velocity of sound through air can then be regarded as having a constant value subject only to temperature and atmospheric pressure.

Effect of Loudness on the Velocity of Sound

Under abnormal conditions an extremely loud sound will have a higher-than-normal velocity, but this holds true only for an extremely short time, when the sound field first leaves the source. The velocity of the sound then assumes the value of velocity measured for lower sound levels.

Effect of Temperature on Sound Velocity

The effect of temperature is to increase the velocity of sound by about 1.1 ft/s for every degree rise measured on the Fahrenheit scale and approximately 0.6 m/s for each degree Celsius.

Temperature affects the velocity of sound in liquids and solids in different ways. In liquids, sound velocity increases as the temperature of the liquid rises, but an opposite effect takes place within solids. As a general rule, sound velocity decreases with rises in temperature in solid materials.

Changes in temperature affect the pitch of sound produced by various musical instruments. For wind instruments, the pitch increases as the temperature rises. The pitch of reeds, bells, and the tuning fork varies inversely with temperature.

For string instruments, the pitch is affected not only by temperature variations but also by changes in string tension due to expansion and contraction.

Effect of Wind on Sound Velocity

An adverse wind can also have an effect on the perception of sound. If the velocity of the sound is such that the sound moves in the same direction as the wind, there is the possibility of greater sound travel. If the sound field travels in a direction opposite that of the wind, more sound energy will be dissipated in less distance.

Table 3-17 gives velocities for various liquids and solids.

TABLE 3-17 VELOCITY OF SOUND IN LIQUIDS AND SOLIDS

| | Sound Velocity | |
Material	(ft/s)	(m/s)
Alcohol	4,724	1,440
Aluminum	20,407	6,220
Brass	14,530	4,430
Copper	15,157	4,620
Glass	17,716	5,400
Lead	7,972	2,430
Magnesium	17,487	5,330
Mercury	4,790	1,460
Nickel	18,372	5,600
Polystyrene	8,760	2,670
Quartz	18,865	5,750
Steel	20,046	6,110
Water	4,757	1,450
Air	1,130	344

THE OCTAVE

Starting with any audible frequency and doubling it represents an octave, a two-to-one range of frequencies. It is also 12 notes of a musical scale

```
 1 2    3 4    5 6 7
 C C#   D D#   E F F#

   G G#   A A#   B
   8 9    10 11  12
```

The separation between 50 Hz and 100 Hz is one octave; from 100 Hz to 200 Hz is the next octave. Any starting point can be chosen. Thus, from 16 Hz to 16,384 Hz is 10 octaves, as indicated in Table 3-18.

TABLE 3-18   OCTAVES AND FREQUENCY RANGES

Frequency Range, Hz	Octave
16 to 32	First
32 to 64	Second
64 to 128	Third
128 to 256	Fourth
256 to 512	Fifth
512 to 1,024	Sixth
1,024 to 2,048	Seventh
2,048 to 4,096	Eighth
4,096 to 8,192	Ninth
8,192 to 16,384	Tenth

Figure 3-3 is a selected octave on a keyboard.

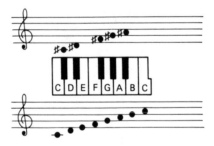

Figure 3-3   One possible octave on the piano keyboard.

4

Microphones

Microphones (mics) belong to the transducer family, devices for converting one form of energy to another, which include components such as batteries, light bulbs, electrical generators, electric heaters, computers, and telephones. As transducers all microphones, regardless of type, change acoustic power into electrical power having essentially similar wave characteristics.

BASIC CHARACTERISTICS

There are various basic characteristics used for identifying microphones. The first of these is the type of moving element—that is, the part that moves in response to sound pressure. Another is the type of pickup pattern, whether the mic responds to sound from all directions or from a limited number. A mic may be specified in terms of its electrical or electronic characteristics. Some microphones have special features and so are given special names.

The output of a microphone is in the form of an AC voltage, varying at an audio frequency. As such, it is amplified, possibly modified in some way, and then reproduced by headphones, speakers, or both.

Basically, a microphone is a simple device. Except for the carbon and condenser microphones, it requires no outside source of power. Its simplicity is evidenced by the fact that it is not equipped with light-emitting devices (LEDs), has no meters, and—with the exception of some specialized units—has no operating controls. Until the advent of synthesizers, microphones represented the start of all sound sources. Components that follow the microphone, such as pre- and power amplifiers, can alter sound, but they cannot improve it. Thus, sound quality is directly dependent on the microphone.

The Electroacoustic Transducer

All microphones are transducers, converting one form of energy to another. The input energy is mechanical; the output energy is electrical. In energy conversions there is always some loss—that is, no transducer has an efficiency of 100%. The loss is in the form of heat, but this is so small that it is not perceptible to human touch.

Because the input is mechanical and the output is electrical, microphones form a separate family of electroacoustic transducers. These can be classified as follows:

1. Controlled, variable-resistance types. This group includes carbon mics.
2. Piezoelectric. This group includes ceramic and crystal mics.
3. Electrodynamic. Dynamic (moving coil) mics belong under this heading.
4. Electrostatic. This heading includes condenser and electret microphones.
5. Electromagnetic. These extremely tiny mics are intended for use in hearing aids.

MICROPHONE TYPES

The Carbon Microphone

The oldest and least sophisticated of all microphones, the *carbon microphone* is essentially a variable resistor whose resistance at any moment is controlled by its sound input. Its basic structure, as indicated in Figure 4-1(a), consists of a flexible metallic diaphragm. Pressing against the center of this diaphragm is a button containing granules of carbon. A resistor, identified by R in Figure 4-1(b), is the load and is connected to a DC voltage source.

With this arrangement a steady direct current flows from the voltage source to the diaphragm, to and through the carbon granules, through the load, and then back to the battery. A sound impinging on the diaphragm causes it to vibrate, altering the pressure on the carbon granules and changing their total resistance. The resulting varying current flows through the load, producing a variable voltage across it. This voltage corresponds in frequency and amplitude to the changing air pressure generated by the sound on the front face of the diaphragm and represents the conversion of acoustic changes to an alternating voltage.

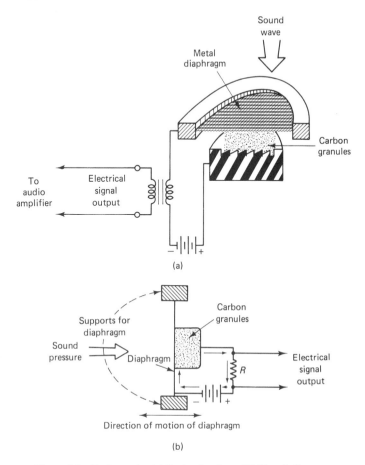

Figure 4-1 Carbon mic. (a) Basic structure. (b) Circuit diagram.

Resistor R can be a fixed-value unit but can also be variable, since the signal voltage output is high compared to other mics.

The primary winding of an audio step-up transformer can be used instead of a load resistor. The secondary winding of the transformer could be connected to the input of an audio amplifier, although this is generally not necessary because of the mic's high sensitivity. The most common application is in connection with telephones, although other microphone types are being substituted in that application. It is not used for hi-fi sound systems because of its limited frequency coverage, noise level, and distortion.

Output level.   The signal-output voltage of the carbon microphone can be calculated from

$$E_0 = \frac{eR_L}{(R_m + R_L) + (hx \sin \omega t)}$$

where E_0 is the AC component of the output voltage, h is a constant in ohms per centimeter, x is the amplitude of the diaphragm in centimeters measured at its point of maximum displacement, ω is equal to $2\pi f$, with f measured in hertz, R_m is the resistance of the mic, e is the voltage measured across the load, R_L is the resistance of the load in ohms, and t is the time in seconds.

Advantages and disadvantages.   The response pattern is omnidirectional and so the unit is capable of picking up unwanted noise from areas to the side and back of the microphone. The microphone is used mainly for voice transmission, since it has limited frequency response. It has a high noise level compared to other microphones and consists mainly of high-frequency hiss due to contact resistance between the carbon granules. Still another disadvantage is packing of the granules when the microphone is subjected to shock, a condition that can also result from excessive current flow.

Carbon mic characteristics.   The carbon mic has a limited pickup range. Its output impedance is low, a relative term, since the actual impedance in ohms of any mic is usually not specified. It is commonly found in communications systems, is quite rugged, and can be used in virtually any environment.

Double-Button Carbon Microphone

The *double-button carbon microphone* has a close resemblance to the single-button carbon microphone, except, as its name indicates, it uses a pair of buttons. The buttons are positioned facing the center of the diaphragm, with one button on each side. Acoustic pressure waves striking the diaphragm cause it to exert a varying pressure on the diaphragm, with more pressure against one button and less against the other. Since the diaphragm vibrates, each of the buttons, in turn, receives more or less pressure.

The advantage of this microphone is that it supplies a more uniform frequency response than the single-button type, but the unit is still plagued by hiss plus a tendency of the carbon granules to compact when subjected to physical shock.

Piezoelectric Microphone

Also known as a crystal microphone, the *piezoelectric microphone* at one time used Rochelle salts as its piezoelectric element. Piezoelectric substances are transducers, converting mechanical pressure to a voltage. Rochelle salts are subject to the effects of humidity and heat and so have been replaced by ceramics such as lead zirconate and barium titanate, substances that are much less sensitive to the environment.

As in the case of carbon microphones, the piezoelectric has a diaphragm that is subjected to acoustic pressure. A drive pin is attached to the center of the diaphragm on the side opposite that receiving sound waves. The crystal element is in the form of a sandwich, consisting of a pair of metal plates with crystal placed between them. Sound waves activate the diaphragm, which pushes the drive pin against one of the metal plates, producing a greater or lesser pressure against the crystal.

This pressure deforms the crystal, with the crystal restored to its original shape when the sound pressure ceases. This double action, crystal deformation and restoration, produces an alternating voltage at the rate of sound pressure change. The ability of the crystal to generate a voltage is known as *piezoelectric effect*.

Characteristics of the crystal mic. The crystal microphone, unlike the carbon, does not require a DC source voltage. The diaphragm is made of a thin section of aluminum, although paper or a film made of polyester plastic can be used. An advantage is its moderately high signal output, so it can be used directly to drive an amplifier having a high input impedance. The unit is not environmentally sensitive, is low in cost, and has a frequency response that can reach 10 kHz.

Prior to the advent of solid-state electronics, the ceramic microphone was very popular, especially with low-cost amplifiers and where high-fidelity sound was not a factor. The microphone using Rochelle salts was referred to as a crystal; those using ceramic materials were called ceramic, although both types, crystal and ceramic used the piezoelectric effect as a working basis.

Dynamic Microphones

The *dynamic microphone*, also called an electrodynamic or moving-coil microphone, shown in Figure 4-2, is somewhat similar to a loudspeaker with the exception that the microphone is a generator; the speaker a sound reproducer.

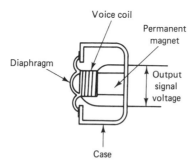

Figure 4-2 Dynamic mic.

In its construction the dynamic microphone has a coil of wire, consisting of a few turns, attached to the center of a diaphragm made of some plastic material, with the diaphragm suspended around its circumference. Sound impinging on the diaphragm moves it back and forth along one arm of a strong permanent magnet. This movement induces a variable voltage in the coil, with that voltage corresponding to the frequency and amplitude of the varying sound pressure. This signal output is taken from the two end leads of the coil.

Advantages of the dynamic mic.   The dynamic microphone is one of the most popular and widely used types. It has one of the most rugged designs, can be used indoors or out, and is capable of having a smooth, wide frequency response. It has good transient response, is highly reliable, and has moderate cost. It is widely used in studio and home recording.

The housing of the dynamic microphone (Figure 4-3) can be modified to produce desired response characteristics. The techniques used in the manufacture of bass reflex speakers are applicable for the extension and improvement of low-frequency response.

In this microphone, air that is trapped at the rear of the microphone housing is allowed to escape through the front via one or more vents. This technique supplies reinforcement to bass sounds. In some instances the microphone is equipped with a bass attenuation control so that the bass-level response can be adjusted. This is a mechanical operation and is done by gradually closing an internal port using a rotating ring positioned below the head. It can also be done electronically by a tone control.

Although dynamic mics may look alike externally, the mass, dimensions, shape, and efficiency of all its parts plus maximum utilization of the field surrounding the permanent magnet all have an effect on the response of the mic, including the amount of distortion and accuracy of translation of the acoustic input into a corresponding signal voltage.

In a representative dynamic mic, the diaphragm is made of a polyester film having a thickness of approximately 0.0015 in. and a diameter of about $\frac{7}{8}$ in. The flux density in the gap between the mic coil and its permanent magnet is typically 10 Gauss or 1 weber/m. The coil fastened to the diaphragm is self-supporting and uses extremely fine wire in its construction, such as No. 48 to No. 50 AWG. Four layers of wire are used, but the entire assembly is very light.

Dynamic mic voice-coil voltage.   The voltage developed across the voice coil is

$$E = BlV$$

where E is the AC voltage across the coil, B is the flux density in the gap between the windings of the coil and the permanent magnet measured in webers per meter, l is the length of the conductor in the air gap in meters, and V is the velocity of the coil in meters per second.

Characteristics of the dynamic mic.   The AC impedance ranges from 150 to 250 Ω, so this mic is a low-impedance type. A typical frequency response is 60 Hz to 15 kHz, but in others the range is from 80 Hz to 18 kHz.

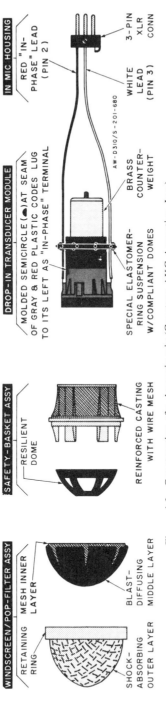

Figure 4-3 Construction of a dynamic mic (Courtesy AKG Acoustics, Inc.)

WINDSCREEN/POP-FILTER ASSY

RETAINING RING

MESH INNER LAYER

BLAST-DIFFUSING MIDDLE LAYER

SHOCK-ABSORBING OUTER LAYER

SAFETY-BASKET ASSY

RESILIENT DOME

REINFORCED CASTING WITH WIRE MESH

DROP-IN TRANSDUCER MODULE

MOLDED SEMICIRCLE (●) AT SEAM OF GRAY & RED PLASTIC CODES LUG TO ITS LEFT AS "IN-PHASE" TERMINAL

BRASS COUNTER-WEIGHT

SPECIAL ELASTOMER-RING SUSPENSION W/COMPLIANT DOMES

AW-D310/S-201-680

IN MIC HOUSING

RED "IN-PHASE" LEAD (PIN 2)

WHITE LEAD (PIN 3)

3-PIN XLR CONN

Figure 4-4 shows the wiring diagram of a moving-coil dynamic mic. The sensitivity can be -60 dB (0 dB $= 1$ mW/10 dyn/cm^2). The impedance for a representative dynamic mic is 600 Ω nominal and can match 150- to 1,000-Ω inputs. For connection to a high-impedance input, a matching transformer can be used, with the transformer located as closely as possible to a following mixer or amplifier input. Long cable runs from the mic to the transformer should be low impedance for minimum noise and hum.

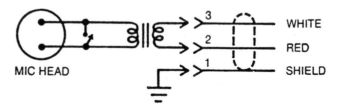

MIC HEAD

3 — WHITE
2 — RED
1 — SHIELD

Figure 4-4 Wiring of the dynamic mic (Courtesy Audio-Technica, U.S., Inc.).

Double-Element Dynamic Microphone

A *double-element dynamic microphone*, also referred to as a coaxial, or two-way, follows the techniques used in coaxial speakers, in which one speaker, such as a tweeter, is mounted coaxially in front of another, typically a midrange/woofer. In the microphone a pair of coils is similarly mounted, with the audio frequency crossing over electrically and acoustically at some low frequency such as 500 Hz.

Ribbon Microphone

Also known as a velocity microphone (Figure 4-5), the *ribbon microphone* consists of a thin stretched duraluminum ribbon about $\frac{1}{4}$ in. wide and 2 to 4 in. long, which,

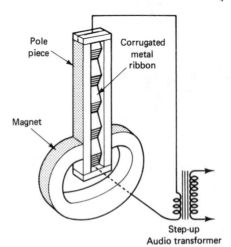

Pole piece
Corrugated metal ribbon
Magnet
Step-up Audio transformer

Figure 4-5 Structure of a ribbon mic.

like the coil of a dynamic microphone, is free to move while positioned between the poles of a strong permanent magnet. The length of the ribbon is the moving portion, but its ends are clamped.

Basically, the dynamic and ribbon microphones work in the same manner and consist of a metallic element—a coil in the dynamic microphone and a ribbon in the ribbon microphone—moving in a fixed magnetic field. A voltage is induced in the coil or ribbon when it moves in this field.

The inductance of the ribbon is extremely low and so is its DC resistance, with the latter about 1 Ω. Because this resistance is so much lower than its inductive reactance, it can be considered as equivalent to the impedance. A step-up transformer is often included in the microphone case, working as an impedance transformation device and enabling a more reasonable impedance match to the input of an amplifier. Another advantage of the transformer is the voltage step-up it supplies.

The ribbon microphone transformer has an impedance step-up ratio of approximately 200:1, so the actual output impedance can be regarded as 200 Ω. With the transformer built into the case, the microphone is considered a low-impedance type. But while 200 Ω is a typical impedance value, these microphones are available with impedances of 30, 150, and 250 Ω.

One of its electrical characteristics is its rather wide frequency response, extending from 35 Hz to 18 kHz, plus or minus 4 dB, although some microphones in this category do have a wider response.

The flat portion of the ribbon faces the front and rear of the microphone, whereas its edges face the sides. Consequently the microphone is sensitive to sounds coming at it from the front and back but has very little response to sounds generated at its sides. The plotted response is a bidirectional, or figure-8, pattern, characterized by insensitivity to sounds that are 90° off-axis.

Physical and electrical characteristics of ribbon mics.   Ribbon mics have an extended frequency response, with some having excellent characteristics in the treble range; the response tends toward flatness over the entire audio spectrum. The output voltage is substantially lower than that of the dynamic. The treble roll-off is gentle, and these microphones do not have much of a popping problem. Transient response is excellent because of the low inertia of the ribbon.

Condenser Microphones

The word *condenser* is a holdout from the time when all capacitors were called condensers. Except for this microphone, condensers are no longer referred to as such in electronics.

The basic structure of this microphone can be represented by a capacitor. One of the capacitor plates is fixed; the other is movable.

An increased current flow into a capacitor depends on a number of factors: an increase in the applied voltage, an increase in the area or the number of plates of the capacitor, and the amount of separation of the plates. Current flow out of the capacitor can be due to the reverse of these factors. Thus by making one of the plates movable, the outward current flow can be controlled. If acoustic energy is applied to the movable plate, the outward current flow can be an audio current whose frequency and amplitude are proportional to the audio energy input.

The movable plate is a metal plate diaphragm, stretched tightly but capable of motion. The diaphragm has a very low mass, essential for the reproduction of transient sounds, and is generally constructed of a plastic substance such as polyester film. To make it conductive, it is given a very fine coating of gold. This coating forms one of the plates of the condenser; the plastic film becomes the dielectric. In the construction of the microphone, the dielectric faces a gold-covered ceramic back plate. Essentially, then, what we have is a variable capacitor whose charge voltage, at any moment, is controlled by the acoustic energy input. The capacitor is given an initial charge, and it is this initial charge that is changed, either to a greater or lesser amount, by the sound input. This initial charging voltage, known as a polarizing voltage, can be supplied by a battery either mounted inside the microphone case or used externally. The amount of voltage that can be impressed on the capacitor depends on its capacitance; that, in turn, depends on its construction.

The output voltage consists of a plus or minus variation of the polarizing voltage, but this variation is quite small, so much so that it cannot be used directly. The output impedance is extremely high, making it impractical to connect a cable directly. Instead, an impedance converter, built into the microphone, works as a connecting link between the microphone and the input impedance of the following component, an amplifier. This unit and the polarizing battery are contained inside the microphone housing. Although the amplifier boosts the signal, it also works as an impedance step-down device, lowering it to some value between 50 and 200 Ω.

Phantom power. The electrical power for the condenser microphone can be supplied by a battery, or the microphone can use power taken from an external device, such as an amplifier or recording equipment. This latter technique is referred to as *phantom powering* or simplex. The signal output cable can be a four-wire device, with two used for conducting the signal out of the microphone and another pair used for supplying the microphone with input power.

Figure 4-6 shows a circuit diagram of a phantom powering unit. The upper drawing in Figure 4-7 supplies the dimensions of a condenser mic. The lower drawing is the circuit diagram. It includes the DC-to-DC converter and power supply.

Characteristics of the condenser mic. Because of the very low mass of its diaphragm, this mic has excellent transient response. With the use of the impedance converter, its output impedance is low. At the input to the converter, the impedance is approximately 10 MΩ.

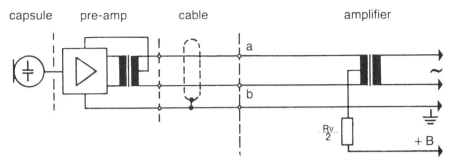

Figure 4-6 Circuit of phantom powering method (Courtesy AKG Acoustics, Inc.)

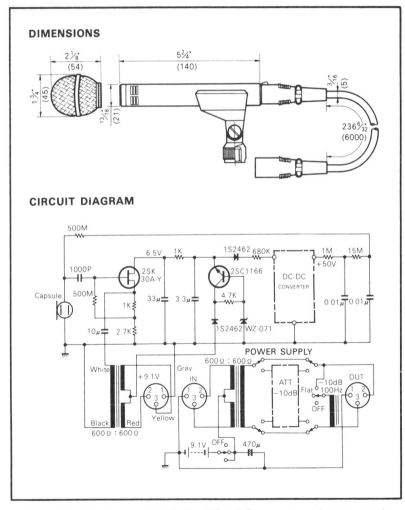

Figure 4-7 Condenser mic including DC-to-DC converter and power supply (Courtesy Nakamichi Co).

The Electret Microphone

The *electret microphone* is a type of condenser mic. The electret is a capacitor-like device capable of holding an electrical charge indefinitely, and its advantage is in the removal of the need for a polarizing voltage. A voltage is impressed across the electret at the time it is manufactured. Known as bias or a bias voltage, it can have a potential of 100 V or more. However, the electret unit can hold only about 20% of the charge placed across it. The compensation is to bring the two plates of the capacitor element closer to each other.

As in the case of condenser mics, the electret type requires an impedance converter generally using a FET transistor. The voltage source for this unit can be a small penlight-type battery. However, some electret microphones do use phantom powering, possibly obtaining DC operating power from a following microphone mixer circuit, with this power supplied by a microphone cable.

The dielectric is fastened to a perforated backplate, which works as the fixed plate. The dielectric is the part of the unit that receives a permanent electric charge. The electret mic has the same advantages as the condenser type. It has a uniform frequency response and good transient response capabilities.

RF Condenser Microphone

The *RF condenser microphone* is a studio condenser mic that does not use a DC bias polarizing voltage. Instead, the condenser element is part of a tuned radio-frequency (RF) circuit operating at 8 MHz. The frequency is generated inside the mic by a crystal-controlled oscillator.

The capacitor is used as the frequency-determining capacitance in a discriminator-tuned circuit. The capacitance variations of this mic capsule due to the sound pressure fluctuations cause the delivery of the audio frequency signal from this discriminator circuit.

BOUNDARY LAYER MICROPHONES

Boundary layer microphones are available under various trade names, such as disc and pressure zone.

Disc Microphone

The *disc microphone* is so named because it looks like a disc with its transducer embedded in a boundary plate. The transducer is inherently insensitive to vibrational noise. This mic's sensitivity and frequency response depend largely on the placement of the disc. It has a sensitivity of 20 mV/Pa, approximately equal to 33 dBV re 1 V/Pa when mounted on a large boundary. The maximum sound pressure for 1% total harmonic distortion is 63 Pa, approximately 130 dB SPL.

The impedance at 1 kHz is 600 Ω or less. The recommended load impedance is 2,000 Ω or more. The polar response is omnidirectional in front of the plate and is hemispherical.

The disc can be mounted on a large surface, such as a wall, floor, ceiling, or piano lid. It will deliver an output voltage 6 dB higher than that of a conventional omnidirectional mic, resulting in a better S/N ratio. The transducer is flush-mounted in the disc, faces forward, and is a self-polarized condenser type. There are no disturbing comb filter effects.

Pressure Zone Microphone

The *pressure zone microphone* (PZM) is an electret type and is positioned in the center of a plate, known as a boundary plate, measuring 5 in. × 6 in. × 1/8 in. thick. The plate has a thin rubber pad in each corner and so is lifted above a surface by about $\frac{1}{32}$ in. The mic itself is positioned about $\frac{1}{32}$ in. above the boundary plate. The mic is powered by a battery or makes use of a phantom supply.

In a conventional mic (Figure 4-8(a)), the unit receives both direct and reflected sounds. When the two sounds are in phase opposition, there will be some cancellation, with the maximum occurring when the two are 180° out of phase. There is also the possibility that the sounds may be in phase, resulting in reinforcement.

With the PZM the direct and reflected sounds arrive at the transducer at practically the same time, so the effect is one of reinforcement. The effect is most

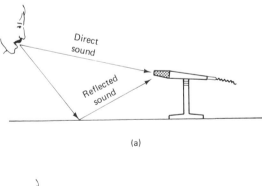

(a)

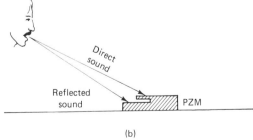

(b)

Figure 4-8 (a) Conventional mic receives both direct and reflected sound. (b) With PZM, separation is much smaller.

noticeable when the mic is used for normal speech. When the audio frequencies are below 350 Hz, a larger boundary panel is required to prevent the drop-off of lower frequencies whose wavelengths are longer than the dimensions of the boundary plate. Figure 4-8(b) shows the short separation between direct and reflected sound when using the PZM. To avoid the drop-off of bass tones, the PZM is mounted on a large, flat surface such as a table top or any other flat surface with an area 2 ft^2 or greater. The PZM has a hemispherical pickup pattern, with the boundary plate functioning as the equator of the hemisphere.

PZMs are designed to be used on flat surfaces such as tables, floors, or walls. Characteristically, they have a smooth frequency response, are free of phase cancellations, and have a good reach (clear pickup of quiet distant sounds), a hemispherical polar pattern, and an uncolored off-axis response. An integral handle, as shown in the photo of Figure 4-9, allows the mic to be hand-held, stand-mounted, or simply put down on any hard surface.

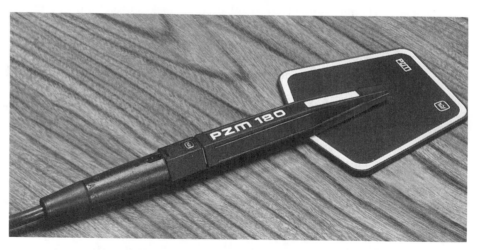

Figure 4-9   PZM positioned on a large, flat surface (Courtesy Crown International, Inc.).

The PZM can be phantom or battery powered. The mic can operate for hundreds of hours on a self-contained N-cell battery. Connecting the mic to a phantom power supply automatically disconnects the internal battery.

The output of the mic is balanced, is low impedance, and uses an XLR connector. Since it is low impedance, the mic can drive long mic cables without hum pickup or high-frequency loss. Although the mic is quite different in outward appearance from other mics, it is compatible with all professional recording and reinforcement equipment. It is equipped with a windscreen to reduce pickup of wind noise and breath pops.

THE SURFACE-MOUNT MICROPHONE

The *surface-mount microphone* is designed specifically for surface-mounting applications. The unit is a permanently biased condenser mic with a hemispherical pickup pattern. This pattern is omnidirectional in the hemisphere above the mounting surface.

The surface-mount mic takes advantage of the principle that at a barrier or boundary, sound pressure doubles compared to its value if the boundary is removed. When placed sufficiently near the boundary surface, the mic effectively has a higher sensitivity of 6 dB and approximately 3 dB greater rejection of random background noise.

Because of its sensitivity, this mic can be used for distant pickup in circumstances where using a mic at close range would not be practical. Since the mic has an omnidirectional polar pattern, sound is picked up equally in the full 360-degree hemisphere around the mic, so there is no off-axis sound coloration or variation.

The surface-mount mic can also be used for close pickup of an individual instrument, mounted, for instance, inside the lid of a grand piano or on the floor next to a bass drum. The unit can be powered by two 9-V alkaline batteries or by a DC voltage simplex supply from broadcast, sound reinforcement, or recording equipment.

Mounting the Mic

To maintain the flattest possible low-frequency response and the best rejection of random background noise, choose a flat surface as large as possible on which to locate the mic. The surface can be a floor, wall, ceiling or table.

A small mounting surface causes a low-frequency roll-off, beginning at the frequency whose wavelength is comparable to the size of the surface. The roll-off continues at a rate of about 3 dB/octave until it reaches a plateau approximately 6 dB lower than the mid- and high-frequency response. In a similar fashion, a small mounting surface decreases the rejection of low-frequency background noise.

Ordinarily, the mic is connected to its following preamplifier by a 7.6-m (25-ft) cable, but up to 15 m (50 ft) of additional cable can be used with no loss in response or output.

Since the mic, like all other condenser mics, is a high-impedance type, it is supplied with a built-in impedance converter. The audio output is via a three-connector balanced cable. The two active leads are color-coded red and blue, and the ground represents the shield braid.

Directional Characteristics

Directional characteristics, called the *polar response*, indicate the microphone's sensitivity to sound pressure from every angle of incidence. Whether a mic is directional or not is a measure of its ability to respond to sounds that are off-axis. It is omnidirectional if its output voltage is essentially constant no matter in which direction the head of the mic is pointed. However, if there is a reduction in signal output when a signal is measured off-axis, the microphone is regarded as directional. The amount of signal decrease is the same for an equal number of degrees to the left or right of the 0° to 180° axis.

A polar pattern chart is a visual representation of a microphone's sound-signal pickup pattern. These are the patterns that indicate whether a mic is omnidirectional, cardioid, bidirectional, supercardioid, etc. Because these directional patterns can vary with frequency, the pickup patterns are often shown at a selected number of frequencies.

POLAR PATTERNS

There are three main polar patterns and a number of variations based on these three.

Omnidirectional.   An *omnidirectional mic* is uniformly sensitive with no preference to sounds from any direction. These mics should be used only when the ambience is an essential part of the recording or where reverberation or acoustical feedback is no problem.

Unidirectional.   *Unidirectional mics* include the cardioid and its variations, the hypercardioid and supercardioid types. The cardioid is most sensitive to sounds coming from the front—that is, on axis (0°)—and less sensitive to sounds from the rear. The shape of the polar pattern resembles a heart, or cardioid. This is the type of mic characteristic most commonly used in both recording and live performances where ambient noise should be suppressed.

Unidirectional patterns are used to eliminate excessive room reverberation, to emphasize a desired instrument or voice, and to reduce feedback when loudspeakers are used in the same room. A variant, the hypercardioid, may be of advantage on stage where feedback is usually a problem.

Bidirectional, or figure-8.   A *bidirectional mic* is sensitive to sounds coming from the front and rear but relatively insensitive to sounds coming from the sides. This characteristic is used for rather special recording cases where rejection of 90° off-axis sounds is important.

Air pressure surrounds mics from all directions, but the mic may or may not be uniform in its response. To describe the sensitivity of a mic, an imaginary line, perpendicular to the head of the mic, is projected. Sound signals directed at the head of the mic are on-axis; those at some angle are off-axis. The line perpendicular to the head is established as a reference and is considered as 0°. The response of a mic can be graphed by first establishing four lines at right angles, starting with the zero reference, continuing to 45° on either side of the reference, and terminating at 180°, actually a continuation of the 0° line. The graph of the response of a specific mic is its polar diagram. Figure 4-10 shows the basic diagram, consisting of a series of concentric circles. Moving inward, each circle represents a decrease in auditory strength of 5 dB and is so marked on the diagram. The mic is shown positioned at the center of the graph with the head of the unit facing 0° and the rear facing 180°.

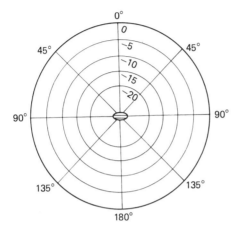

Figure 4-10   Basic polar diagram.

    The advantage of the polar diagram is that it reveals the directional response of a mic. The problem with the diagram is that it is two-dimensional, whereas the surrounding sound pattern is in three dimensions, as indicated in Figure 4-11. The two-dimensional form is more convenient and practical.

Plotting a Polar Pattern

To plot a polar pattern, a sound generator delivering an audio signal, starting at some selected frequency such as 1 kHz, is kept in a fixed, central position, with

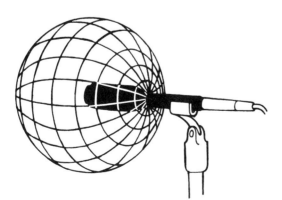

Figure 4-11   Three-dimensional response of omni mic.

the mic rotated around it in a complete circle. The mic is positioned at a distance of 1 m from the sound source, with this distance kept fixed. The signal out of the mic is then indicated on the graph. Although the graph is sectioned into points separated by 45°, plot points can be any selected intermediate values.

The circular movement of the mic can be regarded as a trip along the surface of an imaginary sphere surrounding the mic. A series of such trips can be made, possibly vertical to the mic, horizontal to it, or at any selected angles. Mics are often identified by the polar patterns they produce.

THE OMNIDIRECTIONAL MICROPHONE PATTERN

Figure 4-11 illustrates the three-dimensional omnidirectional pattern of a mic commonly referred to as an omni. Figure 4-12 shows the conventional pattern in two-dimensional form. Tests of this mic were made at three frequencies: 100 Hz, 5 kHz, and 8 kHz. The graph shows that the sensitivity of the mic to sounds arriving from all directions is uniform at a frequency of 100 Hz, but there is some departure from uniformity at 5 kHz and 8 kHz.

Since the omni is responsive to sound from all directions, it will pick up noise from above it, at its rear or front, or from below. It an be used where direct sound from the source plus reflected sound from all directions is wanted. At high

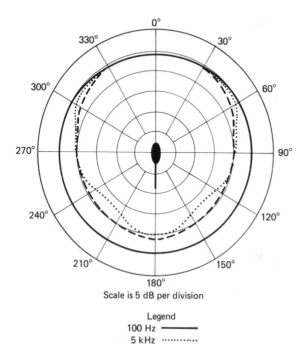

Scale is 5 dB per division

Legend
100 Hz ──────
5 kHz ··········
8 kHz ── ── ──

Figure 4-12 Response of an omni mic at three selected frequencies.

frequencies omni mics show a directional effect. Treble tones that are considerably off-axis, possibly between 90° and 180°, may not have the clarity of on-axis sound.

Omni Construction

The omni mic has a single opening in front of its diaphragm. Consequently, only sound that approaches on-axis is effective. It might seem as though such a structure would make the mic highly directional, but this mic, as all other mics, works in a three-dimensional sound field. As long as that field exists, the mic will supply a corresponding voltage output. This is verified by a test of the unit by having a sound source rotate around it in a horizontal plane, a vertical plane, and at positions at angles to those two planes.

Pressure Gradient

If both the front and rear of a mic diaphragm are exposed to a sound field, a force will be exerted on the diaphragm equal to the sum or the difference between the sound pressures. This resultant pressure is known as a *pressure gradient*. The extent of the pressure gradient is dependent on a number of factors, including the distance of sound-entry orifices located at the front and rear of the diaphragm and commonly referred to as ports, the frequency of the incident sound, and the angle of the sound with reference to the diaphragm. This construction supplies the mic with a directional characteristic. However, if only one side of the diaphragm is exposed to the sound field while the other side is sealed against the environment, the diaphragm will be moved by sound pressure only, with the pressure gradient nonexistent. Under these conditions the microphone will have omnidirectional characteristics.

The Capsule

The capsule in a mic is its transducing element, and in a mic using different sound pressures on both sides of the moving element is sometimes called a *pressure gradient receiver*. Thus, in a cardioid (or unidirectional) mic, both sides of the transducer element have to be exposed to sound pressure.

THE CARDIOID MIC

The cardioid mic is the most commonly used of all the microphones that are specifically made to have a directionality characteristic. Figure 4-13 shows the three-dimensional polar response pattern of the cardioid, so named because of its resemblance to the usual drawing of the human heart. An examination of the two-

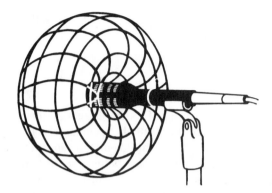

Figure 4-13 Three-dimensional
response of a cardioid mic.

dimensional pattern (Figure 4-14) shows that optimum response is obtained on-axis, with minimum response at the off-axis position—that is, at 180°.

The word cardioid, derived from the Greek, means "heart-shaped." Mathematically, it has the general equation $p = a(1 + \cos\theta)$ in polar coordinates. The response of the cardioid mic is frequency-dependent, with only quality units able to supply uniform frequency response at every angle on- and off-axis.

As indicated earlier, the directional characteristic of the cardioid is achieved by means of external openings and internal passages in the mic that allow sound to reach both sides of the mic's diaphragm. Sound that is 180° off-axis—that is, sound arriving at the rear of the mic—reaches the back of the diaphragm out of phase with the on-axis signal, causing some cancellation. The extent of this cancellation depends on the instantaneous amplitudes of the two signals and whether or not they are completely out of phase.

With the exception of the omni, then, all pattern-generating methods involve phase cancellation to make the mic discriminate between sounds that arrive from the rear and sounds that arrive from the front. They work predictably only when the sources of sound are some distance away from the mic. If you place the mic

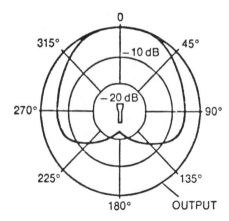

Figure 4-14 Two-dimensional polar pattern of a cardioid mic.

1 in. away from the sound source, the pattern may cost you all the punch you need for mixdown.

The directionality of the cardioid is obtained by taking advantage of the mic's pressure gradient, resulting from ports or holes at the rear of the microphone. Unlike the omni, the head of the cardioid mic, or on-axis position, must be pointed at the sound source. It is the physical construction of this mic that makes it less sensitive to sound reaching it off-axis.

The cardioid is sometimes referred to as unidirectional, implying that it is sensitive only to sound arriving on-axis. What actually occurs is only a reduced sensitivity to off-axis sounds and not to their complete obliteration. The maximum sensitivity reduction is at 180°, but even here there is some sound pickup, depending on its intensity and the distance of the mic from the source.

There are a number of reasons for using a cardioid: (1) to place greater emphasis on the sound supplied by a particular instrument or group of instruments or a vocalist or vocalist group; (2) to minimize background sound, possibly that produced by an audience, or to reduce the sound produced by a noise source; or (3) to reduce or eliminate the possibility of positive feedback when an electronic sound-reinforcement system is used. Such feedback can make the reproduced sound appear to be unusually sharp or, in a worst-case situation, can result in speaker howling.

Three-Dimensional Sound

We call a physical object three-dimensional, since it has three measurable distances—length, width, and height. It may seem strange to refer to sound as such, since it has none of these quantities. However, the reasoning behind speaking of sound as three-dimensional is that sound waves occupy three-dimensional space. But although sound occupies space, the amount of that sound can and does vary throughout.

Figure 4-15 shows the appearance of the response of cardioid, an omni, figure-8, and supercardioid microphones in three-dimensional form. Although this is a more accurate representation of the response of these mics, it isn't as practical as the two-dimensional form. It is the two-dimensional form that is most often used.

A cardioid is not inherently superior to an omni and vice versa. Each has its advantages. The cardioid is more flexible with regard to working distance, the separation between the microphone and a performer or a group. As a general rule, the cardioid can be moved about twice the distance of an omni before encountering excessive reverberant sound, background noise, or feedback from a loudspeaker to the microphone.

If adverse recording conditions are not present, then the omni is probably the microphone of choice, since it may have a smoother frequency response than the cardioid. Omnis are not as susceptible to popping and, depending on con-

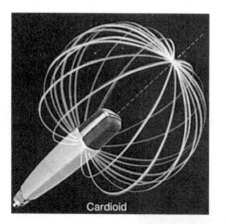
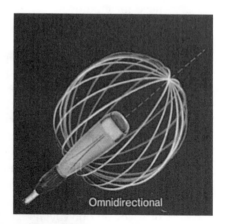
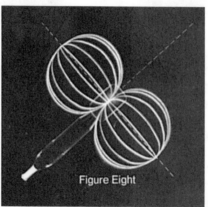
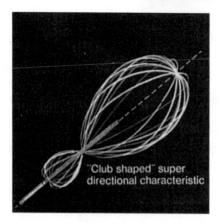
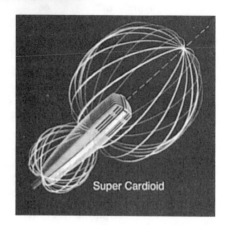

Figure 4-15   Three-dimensional polar patterns of cardioid, omni, figure-8, and supercardioid mics.

struction, may be more resistant to mechanical shock. The cardioid is preferable if audience movements, coughing, and conversation are undesirable.

The signals produced by an omni and cardioid can be very much alike or quite different, depending on working conditions and the intent of the recordist. In a nonreverberant, noise-free environment both types of microphones will sound alike. Where reverberant sound is an excessive part of the acoustical environment, the cardioid would be preferable.

With the cardioid, differences in the frequency response of off-axis sounds will be noted. This characteristic of the cardioid, while to some extent unavoidable, can also be used creatively to suppress unwanted sounds, especially at the higher frequencies.

FREQUENCY-RESPONSE CURVE

The frequency-response curve of a mic provides a picture of the microphone's range and its response within that range. This is just a starting point, because what is recorded in the studio is a complex mix of direct and reverberant sound, the way in which the microphone is positioned, the way it is held (if not mounted in a fixed position), and the acoustics of the studio. This does not mean the frequency response curve is useless, because it is a guide to where the low and high roll-offs occur and to where boosts in response (if any) can occur.

The frequency-response curve represents the microphone's output in decibels and its variations above and below a zero center. The curve does not supply voltage amounts, either absolute or referenced. Although the frequency-range curve may be from slightly below 50 Hz to 20 kHz, the actual range is often well within this amount, with a bottom limit of about 100 Hz and a top limit of 15 kHz. The typical frequency-response curve is that of the mic positioned at 1 m from the sound source. It will be somewhat different for other distances.

The frequency response of a mic is either indicated by a graph over the full audio spectrum or is expressed by two frequency limits in hertz, within which the mic responds uniformly and with little variation of sensitivity. To reproduce music or vocal sound, the mic should (ideally) respond smoothly over the full frequency spectrum. However, as the recording or reproduction environment is not always ideal, it becomes necessary to alter the mic's response to compensate for less-than-perfect situations. For example, a bass roll-off is made to suppress low-frequency resonances encountered in large halls or rumble from air conditioners and other mechanical equipment. A slight boost of the mic's sensitivity at the upper midrange, around 2 kHz to 5 kHz, is incorporated in some mics to emphasize the timbre of certain instruments or voices.

The frequency response of an omni directional microphone is not affected by changing its distance from the sound sources. Any difference in omni fidelity is basically due to the difference in room acoustics introduced by varying the microphone location.

The majority of cardioid microphones do change fidelity when placed closer than approximately 5 ft to certain sources. Such designs are easy to identify. The extra holes in the cardioid mic body required for directional control of sound collection are very close to the head. The end result is an increase in bass response when the mic is placed progressively closer to small sound sources such as a voice, F holes in an acoustic guitar, a cello, a string bass, and similar instruments. The artificial bass boost will decrease as the microphone is moved farther away.

BIDIRECTIONAL MIC

The *bidirectional microphone* is also known as a figure-8 because its polar pattern resembles that digit. The front and rear lobes, as shown in Figure 4-16, are almost identical. The attenuation of acoustic signals is maximum 90° and 270° off-axis, whereas maximum signal pickup is on-axis and at 180°.

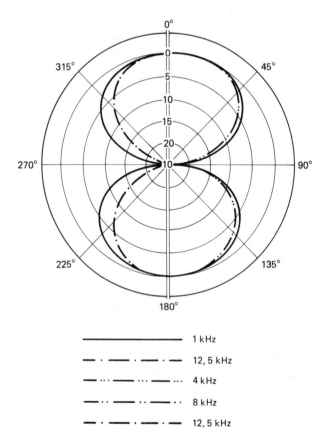

———————————	1 kHz
— · —— · —— · —	12, 5 kHz
— ··· —— ··· —— ···	4 kHz
— ·· —— ·· —— ··	8 kHz
— · —— · —— · —	12, 5 kHz

Figure 4-16 Polar pattern of a bidirectional mic at various frequencies.

SUPERCARDIOID AND HYPERCARDIOID

The response pattern of the supercardioid is somewhat similar to that of the cardioid, but it is different in that it has two response lobes, as indicated in Figure 4-17. One of these is a forward lobe; the other is at the rear. The front lobe is somewhat more elliptical, whereas the rear lobe indicates a greater sensitivity to sounds arriving 180° off-axis. The null regions, areas of little sound response, are shifted to the left and right sides of the mic, with the greatest rejection at 160° and 210° off-axis. The polar pattern of the supercardioid shows that its maximum sensitivity is on-axis but that it is less sensitive to sounds arriving from the sides.

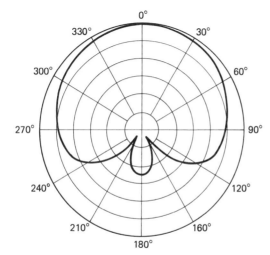

Figure 4-17 Polar pattern of a supercardioid.

The drawing in Figure 4-18 shows the response pattern of the hypercardioid. The rear lobe is more sensitive, but the on-axis response is identical with that of the cardioid. The polar pattern indicates that this microphone is on its way toward being bidirectional. For comparison purposes, the same graph shows the patterns of the cardioid and supercardioid microphones.

The hypercardioid is a dynamic pressure-gradient type with an intrinsically wide frequency range extending from 50 Hz to 20 kHz. The transducer uses two generating elements, a main front-facing sound-pickup transducer and a rear-facing nonacoustic (sealed) noise-compensating transducer. Working together, these transducers reduce the effects of mechanically and motionally induced handling noise by 30 dB at 100 Hz. The main transducer incorporates a hum-bucking winding to cancel the effects of electromagnetically induced noise from power and lighting cables, dimmers, and power switchboards. Its nominal impedance is 370 Ω. In a representative unit its sensitivity at 1 kHz is 1.2 mV/Pa and is −58.4

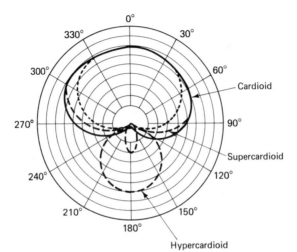

Figure 4-18 Polar patterns of hypercardioid, cardioid, and supercardioid mics.

dBv where 1 Pa $= 10$ dyn/cm$^2 = 94$ dB SPL. Figure 4-19 is a schematic of the hypercardioid.

A hypercardioid has a better sound concentration and a narrower angle of optimum sensitivity than a cardioid. Therefore, a mic with a hypercardioid response pattern picks up less energy put out by a diffused sound field than a cardioid. Sounds that arrive off-axis—that is, from the sides—are even weaker than in the case of the cardioid. This is not due to any peculiarity of the sound field. The hypercardioid is excellent in terms of not being susceptible to feedback.

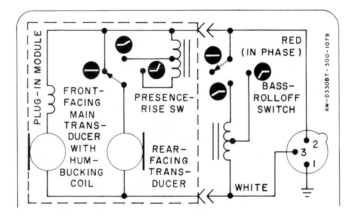

Figure 4-19 Schematic of hypercardioid mic. A positive pressure on the diaphragm of the main transducer produces a positive voltage on the in-phase lead (Courtesy AKG Acoustics, Inc.).

This effect allows higher volume. The hypercardioid also gives better suppression of adjacent sound sources, as, for example, from monitor loudspeakers. The hypercardioid, like the cardioid, is more sensitive to on-axis sounds.

The hypercardioid mic is down only 6 dB at 180°. In some applications, it is desirable to reduce sounds that arrive from a specific, known direction. These mics can be pointed so that the null rejects sound from a specific direction. Thus, pickup of direct sounds from adjacent instruments can be reduced by aiming the mic so the angle of lowest sensitivity points toward these instruments.

The Two-Way Cardioid

The *two-way cardioid* is a dynamic mic that uses two coaxially mounted transducers. One is specifically designed for high frequencies and is placed closest to the front grille; the other is designed for low frequencies and is positioned behind the first. Each transducer incorporates a hum-bucking compensating winding to cancel the effects of stray magnetic fields. Both transducers are coupled to a 500-Hz inductive-capacitive crossover network that is electroacoustically phase-corrected. This is essentially the same design technique used in a two-way speaker system, but applied in reverse.

The frequency range is 20 Hz to beyond 20 kHz. Its nominal impedance is 200 Ω with a recommended load impedance of 500 Ω or higher. Its open-circuit sensitivity at 1 kHz is −77.7 dBv. The maximum power level is −56.5 dBm (re 1 mW/10 dyn/cm^2).

The two-way mic shows performance improvement in the reduction of proximity effect because the high-frequency element keeps the low-frequency element from getting too close to the sound source. Uniform high-frequency response is provided at 90° off-axis, making this mic suitable for wide-angle pickup of vocal or instrumental groups. Cancellation of sound pickup from the rear of the mic is uniform over the entire range of the mic, as compared to many single-element cardioids, which exhibit good back rejection only at middle-range frequencies. In general, the two-way cardioid mic provides 6 dB more amplification before feedback in a sound-reinforcement application than does a single-element cardioid. The wide-area pickup of this mic makes it suitable for podium use where two mics are normally used to capture a wandering lecturer.

THE SHOTGUN MIC

The *shotgun microphone* is closely related to the hypercardioid and has typical cardioid characteristics. It has its greatest sensitivity on-axis and relatively little sensitivity at 90° and 270°.

Its most notable characteristic, though, is its extreme directionality and its reach—that is, its ability to pick up sound sources at long distance when these are on-axis. Because of these operating features, they are rarely selected for studio work but instead are used for on-location recording, especially when the selected site is noisy. They have a high rate of rejection of ambient sound.

PROXIMITY EFFECT

Although not usually shown in frequency-response graphs, some mics have a common characteristic known as *proximity effect*. This is an emphasis of bass response when the mic is used in close proximity to the mouth or other sound source. This emphasis of bass response may sometimes be desirable and is appealing to some vocalists because it allows them to shade their voices. But this same bass emphasis, under other conditions, can lead to a muddy recording. For this reason some mics include switchable bass roll-off, providing the option of either using or neutralizing proximity effect, as desired.

Proximity effect is not encountered in mics with an omni response pattern. With pressure-gradient mics, such as cardioids or bidirectional types, an increase of low frequencies can be noted at distances of less than 2 ft. The pressure difference is related to the velocity of sound, which decreases with the square of the distance in the near field. Since the distance at which the near field becomes a free field depends on the frequency, the near field is larger for low frequencies. This means that a pressure-gradient mic close to the sound source will be in the free field for high frequencies and in the near field for those that are low. Thus the low frequencies will be emphasized.

The most pronounced area in the sound spectrum for proximity effect is below 100 Hz. If the mic, such as a condenser unit, has extended bass response, its low-frequency boost will increase as the frequency is lowered. In hand-held mics where the low-frequency response below 150 Hz is attenuated to minimize handling noise, the proximity effect will produce a noticeable hump around 100 Hz.

The low-frequency boost may be desirable for announcers and vocalists who want an added fullness to their voices. Vocalists may like proximity effect because the increased bass energy produces a high signal-to-ambient-noise ratio, providing greater isolation of vocalists from accompanying instruments, thus giving greater vocal penetration to rock groups and lessened acoustic feedback in some applications. However, for radio announcers who must move around, so that the distance to the mic is constantly changing, proximity effect can cause their on-the-air voices to change character.

Figure 4-20 shows the frequency response of a cardioid mic. At a distance of 1 ft or more, the response is fairly flat. If the working distance is reduced to 3 in., the bass response begins to rise. It is possible to have 20 dB more output at 100 Hz when working at 1/4 in. than when at 1 ft.

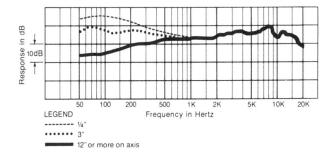

Figure 4-20 Frequency response of a cardioid mic and proximity effect at various distances (Courtesy Audio-Technica, U.S. Inc.).

MICROPHONE DESIGNATIONS

A microphone can be specified in a number of ways: by its polar pattern (such as omni or cardioid), by some feature of its physical structure (dynamic, ribbon, electret) or by some description about the way in which it is to be used (hand held, boom mounted). In some instances combinations are used (boom mounted dynamic omni).

Lavalier Microphone

The *lavalier*, a very small mic, is positioned on the clothing, possibly on a suit lapel or necktie, held in position by a spring loaded clip or suspended from the neck by a string. Its unobtrusive appearance make it ideal for film, television, lecture-hall, and similar applications. The mic may be supplied complete with a built-in IC preamplifier and about 4 ft of nondetachable cable plus a concealed powering module. The lavalier permits performers to have a hands-free delivery and relieves them of being microphone conscious, but it can result in clothing or cable noise, hollow-sounding response, and off-axis depreciation.

Differential (Noise-Canceling) Microphone

The *differential microphone* is designed for close-mouth working in noisy environments. It is intended to be used when the sound source almost touches the microphone's grille. Its design is such that the higher sound pressure on the front of its diaphragm produces electrical output with noise acoustics producing pressure on the other side of the diaphragm to help cancel output. The microphone is insensitive to side-produced sounds and rejects distant sounds in favor of those directed at the front of the diaphragm.

Stereo Condenser Microphone

Separate microphones having similar polar characteristics can be used to supply stereo output. The best arrangement is to use identical models of condenser or electret types. A minimum of two mics is required, but supplementary mics may be required for vocalists or instrumentalists.

Loading

Low-impedance mics have an impedance of 200 Ω or less. Since mics are matched for optimum voltage transfer, the loading of such mics should be as little as possible—that is, the impedance connected to the mic should be substantially higher than that of the mic. As a general rule, the load should be at least three times the mic's impedance. Thus, for a mic having an impedance of 200 Ω, the load should be a minimum of 600 Ω.

Distortion in a sound-reinforcement system and during recording is too often blamed on the mic. A speaker with lips virtually touching the mic or an orchestra can produce up to 100 dB SPL. Loudness impacts on a mic diaphragm held a few inches from screaming vocals, and loud or amplified instruments may reach 120 dB average to 130 dB peak. 130-dB loudness is the threshold of pain.

A good mic will not overload at these pressures. Mic distortion can be held to 0.5% or less with sound pressure up to 130 dB. Some professional mics can be used in the mouth of a trumpet, which can produce up to 146 dB loudness. The mic output remains clean but can become high enough to overload its following amplifier input stage.

The input sensitivity rating of an amplifier or recorder describes the absolute maximum amount of voltage or power it can handle before it overloads and distorts the signal. A comparison of various input sensitivities will reveal a surprising variation. Some amplifiers may be rated to handle a −60-dB input signal before overloading. Others can handle up to −22-dB output from the mic without distortion. There is no standard and there cannot be, since mics are never used at the same distance. Further, sound pressure sources have a tremendous variety and different levels of loudness.

The Shorted Microphone

A mic not equipped with an amplifier will not be damaged by a short across its output terminals. However, its frequency response will be impaired if the mic has to work into a load lower than the minimum load impedance prescribed for the mic. Dynamic mics have a lower limit with regard to their load impedance.

Free-Field Response

The *free-field response* of a mic is defined as the open-circuit voltage (measured in millivolts) generated by the mic at its accessible terminals per undisturbed sound pressure in dynes per square centimeter at a specified frequency.

The measurement is taken in a progressive-plane sound wave at the mic position prior to the introduction of the mic, the mic being placed at a specified angle with respect to the wavefront. It is now more customary to equate the output voltage to $1 \text{ N/m}^2 = 1$ Pa instead of to 1 dyn/cm^2. The conversion is simple, since 1 mV/dyn/cm^2 is equal to 10 mV/Pa.

The free-field response value has little practical significance unless it is accompanied by a statement of the electrical impedance of the mic.

Balanced Output Microphones

The output of a mic can be balanced or unbalanced. An unbalanced output makes use of a shielded cable having a central conductor and a wraparound braid, which works simultaneously as an electrical shield and as a signal conductor.

A balanced arrangement is one in which a pair of individual wires are used as signal conductors. As in the case of unbalanced cable, there is a flexible metallic shield braid; this is not used as a conductor but simply as grounded shield against electrical-noise pickup.

One of the advantages of the balanced arrangement is that the signal currents flow in opposite directions, resulting in some noise cancellation. It is possible to match the balanced output of a mic to the unbalanced input of a following stage, such as a preamplifier, by using a balanced-to-unbalanced matching transformer. There are two possible locations for this transformer: at the output of the mic or at the other end of the run of cable—that is, at the input to the preamp. The preferred location is at the input to the preamp. The reason for this is that using balanced cable results in a smaller possibility of noise pickup. As a result, a lower level of noise is fed to the input of the preamp. All low-impedance mics use balanced output.

Equivalent Noise

A mic in a completely quiet room will still generate some residual noise. With dynamic mics, the noise will be caused by the thermal movements of electrons in the voice coil. Condenser mics have several different noise sources.

The mic's generated noise can be measured using a peak-measuring instrument and a weighting filter (DIN standard 45 405) to put the right emphasis on the most disturbing components of the composite noise. The weighting network is an equalizer with a frequency response that simulates the response of the human ear. It attenuates some sounds and enables the measuring instrument to give an indication of the subjective effect of the noise. The DIN standard requires the use of a peak-reading measuring instrument to follow the weighting network. The *equivalent noise* is then figured by taking the mic's sensitivity into consideration and relating the value to the hearing threshold of 2×10^{-4} μbar. The equivalent noise figure permits a comparison of different mics directly without regard to their sensitivity. A more common way of evaluating the noise characteristics of a mic is by stating the signal-to-noise ratio.

EIA Sensitivity

Depending on its distance from a sound source, a mic will produce a measurable output. If another mic is directly substituted for the first one at exactly the same distance from the sound source and with no change in the signal from that source, the output may be different.

The *EIA sensitivity* supplies a standard figure with all measurements made similarly to allow evaluation of the relative sensitivities of various mics. The measured output is preceded by a minus sign. The higher the number following the sign, the less sensitive the mic. A mic with an output of -55 dB is more sensitive than one with an output of -60 dB.

The sensitivity of a mic is usually expressed as an output voltage per unit of sound pressure at the mic diaphragm. Sensitivity values may be given in terms of Mv/μb, mV/Pa, or dBv. These values are also related to mic impedance; that is, the higher the impedance of the mic, the higher is its sensitivity figure. Both extremes—either too-high or too-low sensitivity—can cause problems when the mic is connected to other sound equipment, such as sound mixers, tape recorders, or preamplifiers. Either overloading or excessive amplifier noise would be the result of such a mismatch between the mic and its associated equipment.

The sensitivity to sound on axis of the microphone (0°) is taken as a reference of 1.0 on its polar diagram. The sensitivity of the mic at any given angle is shown on the polar diagram by the distance of the response curve from the center of the diagram. For an identical input, positioning and frequency, the microphone having a greater output is the more sensitive.

Open-Circuit Voltage

The *open-circuit voltage* is an unloaded figure, that is, there is no voltage drop due to the measuring instrument. The output is supplied in both volts and decibels.

A typical open-circuit voltage for a low-impedance mic could read

$$-80 \text{ dB re 1 V/μbar or } -80 \text{ dBV}$$

This means that for a sound pressure of 1 μbar (74 dB SPL equivalent to the pressure produced by a normal speaking voice 2 or 3 feet away), the unloaded voltage would be

$$-80 \text{ dB with 0 dB} = 1 \text{ V}$$

A less sensitive mic would have a larger negative decibel number, whereas a more sensitive mic would have a smaller negative number.

Impedance

The *impedance* of a microphone is a vector combination of resistance, inductance, and capacitance; in the case of the latter two, it consists of reactance. Of these, however, only the resistance value within the relevant frequency range determines the microphone's basic impedance value, which is expressed either in ohms or some multiple such as kilohms.

Microphones may or may not have built-in voltage amplifiers. As an example, the electret mic has an enclosed amplifier, but its function is more that of an impedance converter than an amplifier, and it is used for reducing the output impedance of this particular mic.

Low-impedance mics, usually 50 to 200 Ω, should feed amplifier input impedances of 500 Ω or higher for minimum degradation over the full-frequency spectrum.

High-impedance mics, usually 50 kΩ should face input impedances of 150 kΩ or higher for best results and no loss of sensitivity.

Old amplifiers with high-impedance inputs sometimes require a higher voltage level from the mic. In this case, a transformer should be used between a low-impedance mic and the amplifier to convert the low-voltage level from the mic to an acceptable level for the amplifier's input stage.

Equivalent Noise Level

The electrical self-noise of a mic is compared to the theoretical output voltage with a sound pressure level at the threshold of hearing (2×10^{-4} μb, or 20 μPa). The measuring unit (dB SPL) is related to an output voltage at 20 μPa sound pressure level.

Hum Sensitivity

External hum fields are produced by various sources, including power cables, light installations, electric motors, and electric tools. Mic sensitivity to such hum fields is expressed as output voltage per induced field strength.

Cable Color Codes, Wiring and Phasing

Following cable color codes and wiring connectors uniformly will usually result in consistent phase relations. It is important to wire all cables exactly alike to ensure interchangeability. As a final check, a Y cord can be made, which permits connecting two mics to a single input with both mics in phase. Speak into either mic and then into both mics held close together. If the mics are wired alike, there should be no change in level or response compared to a single mic (the mics should be similar for this test).

Considerations of phasing do not end with the mic wiring. Where mics are located both with regard to the instruments and to each other will have a strong effect on sound quality. For instance, you might use mics to create a stereo effect with a drum kit by placing an overhead cardioid at either side of the drum setup. It may sound fine when the drums at each end of the kit are played. But drums near the center may sound strangely muffled or change character, depending on where the head of the drum is struck. This is caused by the same sound arriving at both mics at slightly different times due to the different distances of the mics from the sound source. The sounds are thus out of phase at some frequencies but perhaps in phase at other frequencies.

Correct *phasing* is essential for any studio system having more than one active mic. The operating information accompanying the mic will indicate the procedure to follow for correct phasing. The in-phase terminal of a mic is the one that acquires a positive potential with respect to the other when the front of the mic is subjected to a positive sound pressure wave. The in-phase terminal is pin 2 on mics equipped with an XLR connector.

Phasing may seem a small matter, but it can have major consequences when more than one mic is used. Placing two mics wired out of phase so that both pick up the same sound may result in some rather weird cancellations at various frequencies. Even a slight movement of either of the mics or the sound source will change the cancellation effect. Such patterns are vastly reduced when the mics and the system to which the mics are connected are in phase.

Sound Roll-offs and Boosts

Offhand, it might seem desirable for all microphones to have a flat frequency response over as much of the audio range as possible. From a practical viewpoint, variations in frequency response are designed into a microphone to compensate for varying acoustical conditions, environmental problems, or creative needs. Such variations are referred to as *roll-offs* and *boosts*, with the degree of variation expressed in decibels. Thus a microphone with some low-frequency attenuation may be said to roll off (drop) some amount, such as 5 dB, for example, at 50 Hz. This might be done because a mic with an extended bass response would tend to emphasize low-frequency rumble from air conditioners and other mechanical equipment. Similarly, there could be some acoustical condition that would make it desirable to boost treble tones to counteract excessive absorption of those tones or to compensate for low energy of the treble in comparison with the midrange or bass.

5

Filters, Pads, and Attenuators

Filters can be *active* or *passive*. An active filter is associated with an amplifier; a passive filter is not. A passive filter consists of one or more electronic components such as inductors, capacitors, resistors, and crystals, designed to have a specific effect on the entry and passage of a single, small or large group of frequencies. These may be in the audio, radio frequency, or microwave range. The fact that a filter is labeled radio frequency or microwave does not mean it is excluded from audio. A filter might be used to separate unwanted radio frequencies from audio.

From a construction viewpoint, some filters are extremely simple and can consist of nothing more than a single component, possibly just one capacitor. Others are quite complex and may include a large group of components. Some filters are so commonly used that they are readily available in manufactured form; others must be designed. Filter design would occupy a large book of its own. Those shown and described here are the more usual types.

FILTER CLASSIFICATION

There are various ways in which filters can be grouped.

1. By the number of their reactive elements. A filter can have one, two, or more. The simplest type of filter will have a single reactive element, either an inductor or a capacitor.

2. By the number of poles. A pole includes a reactive pair, consisting of an inductor and a capacitor. A 4-pole filter has four pairs of associated reactances.

3. By configuration. Configuration is an arrangement of reactances having a resemblance to the letters T, L, or π.

4. By transmission characteristic. A filter can be a low-pass, high-pass, band-pass, or band-reject type.

5. As m-derived.

6. As constant-k.

7. By designer: Chebyshev, Butterworth, and so on.

8. By the mathematical function from which it is derived, as in the case of the Bessel filter.

9. By function. Filters can be named for the work they do, as power supply, noise reduction, and so on.

10. By the sharpness of roll-off, specified in decibels.

11. By whether they are purely electronic using reactive elements or are made on various piezoelectric substrates, as in the case of surface acoustic wave filters.

12. By the amount of insertion loss, measured in decibels.

13. By whether they are cascaded and the extent of such an arrangement.

14. As a special type, such as the comb filter.

15. By whether the filter is passive or active.

16. As elliptic.

17. As iterative RC.

These designations are not exclusive and are frequently combined to make the description of the filter more specific.

18. As digital.

19. As brickwall.

AUDIO OCTAVES AND DECADES

The division of the audio spectrum into decades can be done as follows:

$$20 \text{ Hz to } \quad 200 \text{ Hz} \quad \text{(first decade)}$$

$$200 \text{ Hz to } \quad 2,000 \text{ Hz} \quad \text{(second decade)}$$

$$2,000 \text{ Hz to } 20,000 \text{ Hz} \quad \text{(third decade)}$$

A more detailed division is obtained by dividing the audio spectrum into octaves:

20 Hz to	40 Hz	(first octave)
40 Hz to	80 Hz	(second octave)
80 Hz to	160 Hz	(third octave)
160 Hz to	320 Hz	(fourth octave)
320 Hz to	640 Hz	(fifth octave)
640 Hz to	1,280 Hz	(sixth octave)
1,280 Hz to	2,560 Hz	(seventh octave)
2,560 Hz to	5,120 Hz	(eighth octave)
5,120 Hz to	10,240 Hz	(ninth octave)
10,240 Hz to	20,480 Hz	(tenth octave)

Using decades, the audio spectrum is divided into three sections; with octaves it is divided into ten sections. For the most part, octaves are used in a comparison of the attenuation in decibels of a filter and its operating frequency. Thus, a decade is the interval between two quantities having a ratio of 10:1.

It is also possible to start at any selected frequency, such as 300 Hz, and move either higher or lower. Thus, for 300 Hz there would be one octave at 150 Hz and another octave at 75 Hz.

An octave is a comparison between two audio frequencies having a ratio of 2:1. It can also be defined as a tone that is 12 full tones above or below another given tone. The slope of a filter can be indicated in terms of decades or in decibels per octave. The slope of the low-pass filter illustrated in Figure 5-1 is given as 12 dB/octave. The cutoff point in this graph is a little more than 2 kHz, whereas the

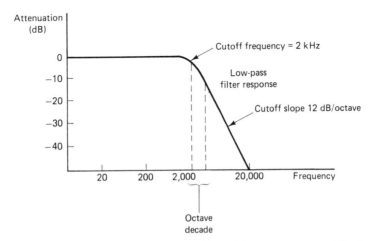

Figure 5-1 Slope of a selected low-pass filter. Other slopes are possible, depending on the design. In this example the slope is 12 dB/octave, or 12 dB/decade.

maximum attenuation is a little more than 20 kHz, or one decade. One octave would be the frequency range from 2 kHz to 4 kHz. For this octave and for succeeding octaves, the slope would be 12 dB/octave.

High-Pass Filter

The simplest *high-pass filter*, shown in Figure 5-2, consists of a single capacitor in series with the line. Since its reactance varies inversely with frequency, it will have its maximum opposition to extremely low frequencies, with that opposition decreasing as the frequency rises. At DC (zero frequency) the opposition is such that the filter works as an open circuit.

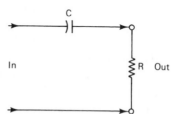

Figure 5-2   Basic high-pass filter.

High-Pass Constant-k Filter

The single-pole filter in Figure 5-3 is known as a *constant-k* type, since the product of its series and shunt impedances is a constant at all frequencies. Thus, in this high-pass filter, the series unit, a capacitor C in series with the line, has a reactance

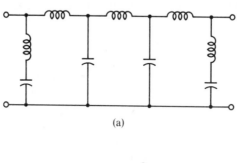

(a)

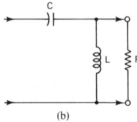

(b)

Figure 5-3   High-pass constant-k filter.

that varies inversely with the applied frequency. The coil, working as a shunt element, has a reactance that varies directly with the applied frequency. As the reactance of one reactive element—either coil or capacitor—increases, the reactance of the other decreases.

The high-pass filter is one in which frequencies higher than a previously selected cutoff frequency (f_c) are passed, whereas frequencies that are lower are attenuated. For a constant-k type:

$$C = \frac{1}{4\pi f_c R}$$

$$L = \frac{R}{4\pi f_c}$$

$$R = \sqrt{\frac{L}{C}}$$

where C is the series capacitance in farads, L is the shunt inductance in henrys, f_c is the cutoff frequency in hertz, and R is the value of the terminating resistor in ohms.

ACTIVE VERSUS PASSIVE FILTERS

An active filter is associated with a signal amplifier and as a result, its output can be greater than its input. A passive filter (Figure 5-4(a)) is not part of amplifier circuitry and results in a signal-voltage decrease referred to as insertion loss. However, the presence of an audio amplifier does not automatically mean signal gain, since the amplifier may be connected as a unity gain follower. For the most part active filters are band-pass types.

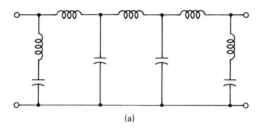

(a)   **Figure 5-4** (a) Passive filter.

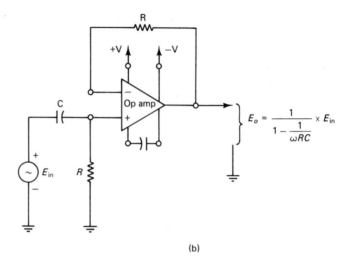

$$E_o = \frac{1}{1 - \dfrac{1}{\omega RC}} \times E_{in}$$

(b)

Figure 5-4 (b) High-pass active filter with roll-off of 20 dB/decade. V is the operating voltage of the op-amp.

OPERATIONAL AMPLIFIER

The operational amplifier, more often referred to as an op amp, is a high-gain, direct-coupled amplifier available in the form of an integrated circuit (IC).

The op-amp has a pair of terminals marked $-$ and $+$. These are referred to as the differential input terminals because the output voltage, E_o in Figure 5-4(b), depends on the voltage difference between them. The input signal voltage is E_{in}; the output signal voltage is E_o. The DC operating voltage is shown as $-V$ and $+V$. The minus signal input $(-)$ is referred to as the inverting input; the plus signal input $(+)$ is the noninverting input. The difference between these signal voltages is indicated as

$$E_o = \text{plus signal voltage input} - \text{minus signal voltage input}$$

Active filters can be designed for a wide frequency range and a high dynamic range and can be made to outperform passive filters. An active filter can have a very high input impedance and so will act as a light load on the signal-supplying source. Unlike passive filters, they can be switched from one type to another.

The active element in the drawing is an IC op amp. The op amp has inputs for a DC voltage and also for the input and output signals. The op amp is a chip containing a number of direct-coupled transistors, associated diodes and resistors. Although often shown in circuit diagrams as a three-terminal device, it can have as many as nine terminals. It can be used as a signal-inverting device, or in a noninverting mode. The output voltage is based on the difference between the input voltages; hence the component is sometimes further identified as a differential mutual transconductance operational amplifier (differential op amp). The op amp uses large amounts of negative feedback and has high stability.

In Figure 5-4(b) the circuit of such an amplifier is built around a high-pass filter. The output voltage is represented by

$$E_o = \frac{1}{1 - \dfrac{1}{\omega RC}} \times E_{in}$$

where E_o is the output voltage, $\omega = 2\pi f_c$, R is the resistance in ohms, and C is the capacitance in farads.

$$\omega_c = \frac{1}{RC} = 2\pi f_c$$

$$R = \frac{1}{\omega_c C} = \frac{1}{2\pi f_c C}$$

The filter is part of an op amp.

ROLL-OFF

The roll-off for this filter is shown in Figure 5-5. The roll-off is the sloping portion of the curve, and it starts at the half-power point. This is the 3-dB point, where the output voltage is 70.7% of its peak value. The slope of the curve is 20 dB per decade. In Figure 5-5 the slope starts at ω_c, with one decade between it and $0.1\omega_c$ and another between $0.1\omega_c$ and $0.01\omega_c$. Each of these decades supplies a slope of 20 dB.

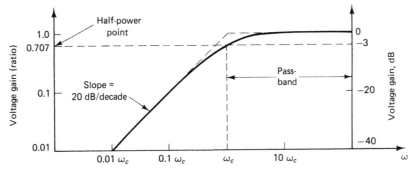

Figure 5-5 Roll-off for high-pass filter.

BASIC LOW-PASS FILTER

Figure 5-6 shows the circuit of a *basic low-pass filter*. It consists of an inductor, L, in series with the line. R is the load resistor. This filter works in a manner opposite that of the high-pass filter. The reactance of the inductor varies directly with the frequency of the input signal. Consequently, low-frequency signals are passed; higher-frequency signals encounter higher reactance.

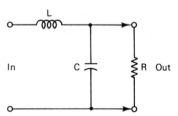

Figure 5-6   Single-pole low-pass filter.

Low-Pass Constant-k Filter

The low-pass filter in Figure 5-6 passes all frequencies below a selected value and attenuates higher frequencies. Its action is exactly the opposite of a high-pass filter, so the circuit arrangement is also exactly opposite. For a low-pass filter:

$$C = \frac{1}{\pi f_c R}$$

$$L = \frac{R}{\pi f_c}$$

$$R = \sqrt{\frac{L}{C}}$$

where C is the capacitance in farads, L is the inductance in henrys, R is the resistance in ohms, and f_c is the cutoff frequency.

Band-Pass Constant-k Filter

This filter whose circuit appears in Figure 5-7 permits the passage of a previously determined band of frequencies while attenuating frequencies below and above the bandpass. The arrangement takes advantage of a series and a parallel LC circuit. These two have different impedance behaviors. A series circuit has its minimum impedance at its resonant frequency; a parallel circuit has its maximum impedance at its resonant frequency.

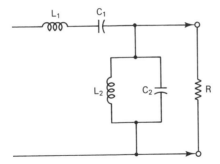

Figure 5-7   Band-pass constant-k filter.

The series arrangement has its minimum impedance at the center frequency of the passband. The impedance increases on either side of this center frequency. The shunt circuit has a behavior that is the opposite. It has its maximum impedance at the center of the passband, with a decreasing impedance on either side. The required values of the components of the filter are supplied by

$$C_1 = \frac{f_2 - f_1}{4\pi f_1 f_2 R}$$

$$C_2 = \frac{1}{\pi(f_2 - f_1)R}$$

$$L_1 = \frac{R}{\pi(f_2 - f_1)}$$

$$L_2 = \frac{(f_2 - f_1)R}{4\pi f_1 f_2}$$

where C_1 and C_2 have values in farads, L_1 and L_2 are in henrys, and R is the load resistance in ohms. Also, f_1 is the lower cutoff frequency of the passband, and f_2 is the upper cutoff frequency.

The disadvantage of this circuit is that for frequencies in the audio range, the inductors and capacitors must be fairly large. The inductors are iron-core types; the capacitors are selected for high capacitance.

Band-Elimination Constant-k Filter

Also known as a band-rejection, band-stop, or band-suppression filter, the band-elimination constant-k filter has a circuit arrangement as shown in Figure 5-8. Like the band-pass filter, it makes use of a series and a parallel tuned LC circuit,

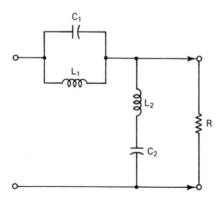

Figure 5-8 Band-elimination constant-k filter.

except that these are now transposed. The values of the individual parts are supplied by

$$C_1 = \frac{1}{4\pi(f_2 - f_1)R}$$

$$C_2 = \frac{f_2 - f_1}{\pi f_1 f_2 R}$$

$$L_1 = \frac{(f_2 - f_1)R}{\pi f_1 f_2}$$

$$L_2 = \frac{R}{4\pi(f_2 - f_1)}$$

where C_1 and C_2 have capacitances specified in farads, L_1 and L_2 are in henrys, R is in ohms, and f_1 and f_2 are the lower and upper cutoff frequencies.

T-Type Low-Pass Constant-k Filter

The circuit shown in Figure 5-9 is referred to as a *T-type filter* because of its fancied resemblance to the letter *T*. Actually, it is just a modification of the low-pass filter shown earlier in Figure 5-6. The effect of this change is to produce a sharper frequency cutoff.

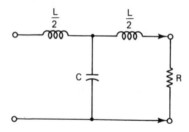

Figure 5-9 T-type low-pass constant-k filter.

π-Type Constant-k Low-Pass Filter

Figure 5-10 shows a variation of the T-type constant-k low-pass filter. The two coils previously identified as L/2 have now been combined into a single inductor. Assuming no coupling between these coils, the replacement coil has a value of $L = L_1 + L_2$. Because of its appearance, the filter is known as a π *type*.

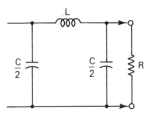

Figure 5-10   Variation of the T-type constant-k low-pass filter.

Multisection T-Type Filters

A high-pass filter can also be made into a T-type by inserting another series capacitor into the line. A *multisection T-type filter*, shown in Figure 5-11, is made by joining two such units and thus has several connected capacitors. Multisection band-pass and band-elimination filters can be made by joining additional sections.

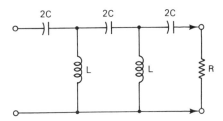

Figure 5-11   Multisection T-type filter.

π-Type Low-Pass Filter

T-type filters are made by putting in additional series elements. A *π-type filter* is obtained by adding another shunt element. As in the method used in T-type filters, multisection units can be made by joining single units. If the two inductors are identical, they can be replaced by a single unit having half the value of either.

Filter Impedance

The impedance of the circuit or component connected across the input of a filter is referred to as the source impedance. The circuit or component across the output of the filter is the load impedance. The filter itself represents an impedance—that is, it has its own input and output impedances. These impedances, which are at each end of the filter, are known as input and output image impedances. For maximum transfer of energy from the filter to the load, the output image impedance should be equal to the load impedance. For maximum input energy transfer, the source impedance should be equal to the input impedance of the filter.

m-Derived Filters

An *m-derived filter* is one that is obtained from one of the constant-k types. These m-derived filters have additional impedances and so have a much sharper cutoff. The type of derived filter obtained depends on the kind of modification made to the basic constant-k type. If an impedance is added to the shunt arm of the basic filter, the result is a series-derived m-type. If an impedance is connected to the series arm, the modified filter is referred to as a shunt-derived m-type. The additional impedances can be coils, capacitors, or series-parallel combinations.

Types of m-derived filters. An m-derived filter can be low-pass, high-pass, or band-pass. Within these three main categories there are single- and multi-element and T sections.

Derivation of m. The ratio of the cutoff frequency f_c to the infinite attenuation frequency f_∞ is a factor that is designated by the letter m. For a low-

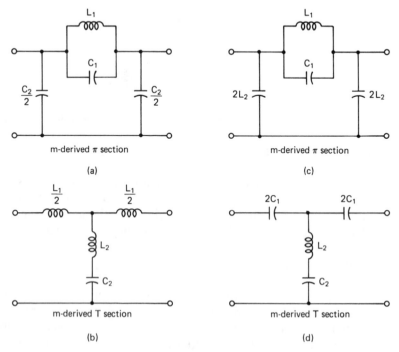

Figure 5-12 m-derived π- and T-type filters.

pass filter the value of m can be derived from

$$m = \sqrt{1 - \left(\frac{f^2}{f_\infty}\right)}$$

and for a high-pass filter,

$$m = \sqrt{1 - \left(\frac{f_\infty^2}{f}\right)}$$

Figure 5-12 shows a number of filters of the m-derived type. The diagram in part (a) is that of a π section; (b) is a T section, (c) is a π section, and (d) is a T section.

FILTER ORDER

Filters are sometimes additionally identified by the number of reactive elements they have. A filter consisting solely of a single capacitor or inductor is a first-order filter. Associated resistors are not included in the count, since these are not reactive elements. Figure 5-13 shows a pair of first-order filters, with one having an inductor and the other having a capacitor as the reactive filter elements. Both of these filters are passive types.

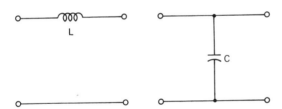

Figure 5-13   First-order filters.

Figure 5-14 illustrates a second-order low-pass filter. The circuit has two capacitors, C1 and C2, and these are the reactive elements that determine the

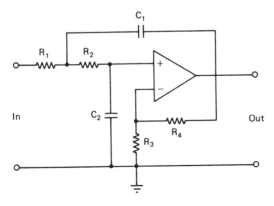

Figure 5-14   Second-order low-pass filter.

order of the filter. There are four resistors, but these do not take part in the filtering action. The filter uses an op amp and so is an active type.

Increasing the Filter Order

The effectiveness of a filter can be increased in a number of ways. One of these is by adding more reactive elements; another is by combining filters into a multiple type; or a third is by using both methods. Such a filter is shown in Figure 5-15. This is a fourth-order filter consisting of a pair of second-order filters.

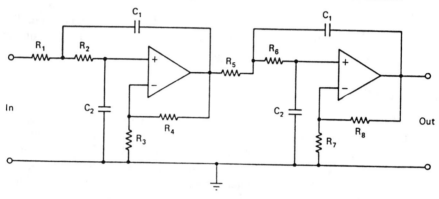

Figure 5-15   Fourth-order low-pass filter.

The graph in Figure 5-16 shows how the effectiveness of a filter is increased as it is moved into a higher-order category. The graph is that of a low-pass filter. The vertical and horizontal dashed lines represent the functioning of an ideal filter. As the order of the filter is increased from 1 to 2 to 4, the filtering action more closely approaches that of the ideal.

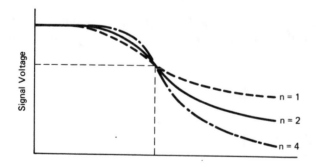

Figure 5-16   The higher the order of the filter (n), the sharper the roll-off.

When a filter contains single reactive elements only, these are considered as tuned elements and so determine the order of the filter. A tuned element, however, may consist of an inductor and a capacitor, either in series or in parallel or combined series-parallel. The number of tuned sections, then, determines the order of the filter. The drawing in Figure 5-17 has four reactive elements, con-

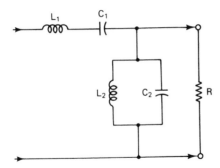

Figure 5-17 Second-order band-pass filter.

sisting of a pair of capacitors and a pair of inductors. However, there are only two tuned circuits, one series and the other parallel, and so this is a second-order filter. Figure 5-18 shows first-, second-, third-, and fourth-order filters using tuned circuits. The higher the order number, the sharper the roll-off. A rectangle represents an ideal filter condition.

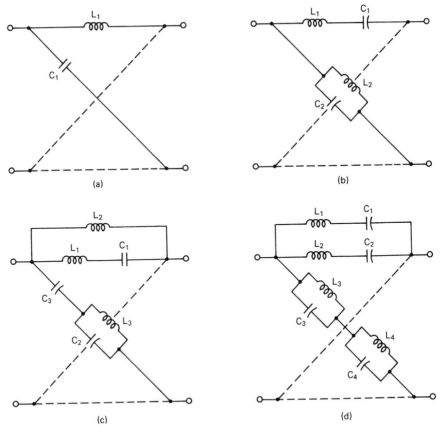

Figure 5-18 Filter order using tuned circuits. (a) First order. (b) Second order. (c) Third order. (d) Fourth order.

It might seem that using a very high order filter would be close to ideal. But the higher the order of the filter, the greater its construction complexity and the greater the need for careful engineering design to make sure the elements of the filter do not interfere with each other or cause other problems. The manufacturing cost is also higher.

MISCELLANEOUS FILTERS

Not all filters are referred to by order number; some are designated, instead, by their function. In some instances any filtering that is done is incidental to the function. Thus, a capacitor used to transfer a signal from one circuit to another is known as a coupling capacitor, and that is its primary function. However, because it is a reactance, it is frequency-sensitive and so it also works as a filter, even though that action may not be wanted.

There are also some reactive circuit arrangements used as filters that are known by function and not by order number. Thus, a filter following the diode rectifier in a power supply may consist of a series inductor and one or two shunt capacitors. It is referred to as a power-supply filter. A similar type of filter may have the same circuitry but has different values of capacitance and inductance. Known as a line-noise filter, this filter is inserted in the AC power line between an AC power source and some electronic component such as a receiver or a compact disc player.

Comb Filters

A *comb filter* is one that has a multiple-band-pass design, passing frequencies within a number of narrow bands. This filter is so named because its response characteristic waveshape resembles the structure of a comb.

SPECIAL FILTER DESIGNS

Butterworth Filters

Butterworth filters are often used as low-pass filters but can be applied in high-pass and band-pass arrangements. They have a passband that is essentially flat, with a rather sharp roll-off. They can be used in either active or passive form. Their operating frequency range is high and extends from DC to RF. The values of R, C, and L used in the filter are dependent on the input and load impedances. The Butterworth filter has a very flat response in its passband. In its stopband its rejection has a constant slope of 6 dB/octave.

Chebyshev Filters

Chebyshev filters, whose name is spelled in various ways—Chevyshev, Tschebyscheff, and Tschebyshev—are quite similar in behavior to the Butterworth types. They can be used in low-pass, high-pass, band-pass and band-rejection configurations, can be used as passive or active types, and have a sharp roll-off characteristic.

Bessel Filters

Also known as linear phase or constant-time-delay filters, *Bessel filters* derive their name from the fact that Bessel functions are used in their design. They do not have as sharp a roll-off as do Butterworth filters. Figure 5-19 is a comparison between Butterworth and Bessel types. Bessel filters are characterized by excellent phase-shift linearity characteristics. These filters are often used in an active configuration in association with an operational amplifier.

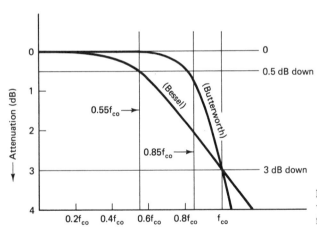

Figure 5-19 Roll-off of Butterworth versus Bessel filter; f_{co} is the cutoff frequency.

Elliptic Filters

Elliptic filters are designed along the lines of Chebyshev filters. Their passband is not as smooth, and, as shown in Figure 5-20, they have substantial ripple in the passband. The amount of attenuation in the rejection band is characterized by fluctuations.

An elliptic function is a mathematical expression and is used in an effort to obtain the squarest possible amplitude response of a filter utilizing the minimum number of filter circuit elements.

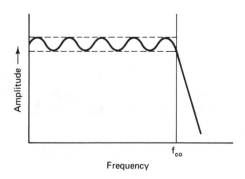

Figure 5-20 Band-pass ripple.

Iterative RC Filters

Iterative RC filters resemble Bessel filters but have a more gentle roll-off. The flatness of their frequency response is also poorer. An iterative filter is a four-terminal type that provides iterative impedances. The word *iterative* indicates repetition.

An iterative impedance, when connected to the input terminals of a transducer, results in the same impedance appearing across the output.

Surface Acoustic Wave Filters

Also known as SAW filters, *surface acoustic wave filters* are small, rugged filters suitable for mounting on printed-circuit (PC) boards and are used to replace conventional filters that use reactive elements such as capacitors and inductors. This filter has a rather high insertion loss and so is used as an active type rather than as a passive arrangement. SAW filters have a wide range of applications, including use in modulators, demodulators, converters, decoders, signal processors, and receivers, including both broadcast and satellite TV.

SAW filters supply excellent amplitude and phase response over a wide range of center frequency and bandwidth requirements. Figure 5-21 shows the band-pass response of one of these filters. The attenuation is very high, in the order of about 60 dB on both sides of the passband.

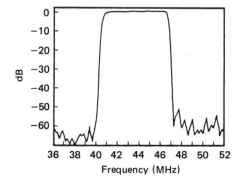

Figure 5-21 Band-pass response of a SAW filter (Courtesy of Crystal Technology Co.).

SAW filters are manufactured on a variety of piezoelectric substrates, the two most common being lithium niobate (LiNbO$_3$) and ST quartz. Quartz demonstrates excellent stability over a wide temperature range, whereas lithium niobate exhibits excellent electromagnetic to acoustic coupling. The name acoustic wave filter is obtained from the fact that minute acoustic waves travel over the surface of the substrate.

Types of SAW filters. There are two major types of SAW filters. The first is a SAW transversal filter. These have excellent passband characteristics and stopband rejection. The second type is a SAW resonator filter used for narrowband, low-loss filters in the range of 30 MHz to 1 GHz or more.

POLES

The number of poles that constitute a filter is determined by the number of associated reactive elements in the filter. This is a significant characteristic, since the roll-off is dependent on the number of poles. Figure 5-22 illustrates a six-pole filter consisting of three filter sections of two poles each. The roll-off for a filter is 6 dB/octave for each pole, as indicated in Figure 5-23. For a two-pole filter the roll-off is 12 dB/octave; for a four-pole type, it is 24 dB/octave.

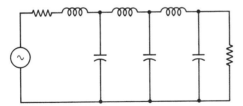

Figure 5-22 Six-pole filter.

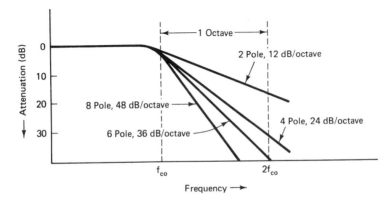

Figure 5-23 The greater the number of poles, the sharper the roll-off.

ANALOG AND DIGITAL FILTERS

One of the possible functions of a filter is its use as a time-delay circuit. Whether a filter is an analog or digital type depends on how it is used. If it is made to work in an analog circuit, it is properly called an analog filter; if in a digital circuit, a digital filter.

The circuit in Figure 5-24 is either an analog or digital type.

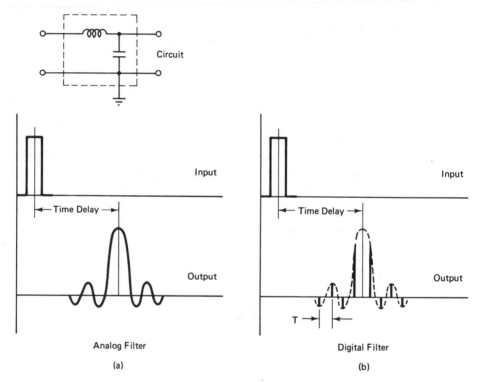

Figure 5-24   (a) Analog filter. (b) Digital filter (Courtesy Onkyo U.S.A. Corporation).

BIQUAD FILTER

An active filter need not contain just a single op amp. The circuit in Figure 5-25, known as a *biquad*, has three such amplifiers. The component is a second-order low-pass filter and contains two reactive components, identified as C_1 and C_2. The advantage of the biquad is that it has good stability, an important factor when several stages are to be cascaded.

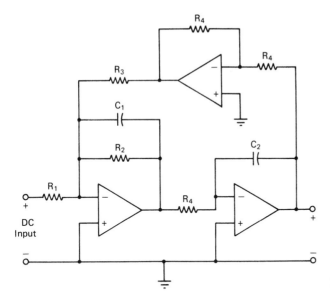

Figure 5-25 Biquad filter.

PADS AND ATTENUATORS

Filters, pads, and attenuators are alike in the sense that they are inserted in series with a line carrying a signal, or shunted across it. Pads and attenuators, however, use only resistors and have no reactive elements. Consequently, they are not frequency-sensitive. Further, they are passive only and introduce a signal loss. This loss is applicable to weak as well as to strong signals.

Pads and attenuators are quite alike except that pads have only fixed resistive elements; attenuators have one or more variable elements. Sometimes, though, the words pad and attenuator are used synonymously.

Pads are identified in the same manner as attenuators, as T pads, H pads, π pads, and so on. The circuit diagrams are the same except that the resistors used for pads are shown as fixed elements. Figure 5-26 shows the circuit arrangements of the three more commonly used pads, T, H, and π. Table 5-1 shows the resistor values for R_1 and R_2 and the insertion loss ranging from 0.1 to 40.0 dB.

Unbalanced versus Symmetrical T Pads

Figure 5-27 shows the arrangements for a pair of T pads. Although the circuits are the same, the resistor values are not. Thus, in the unbalanced unit, the values of the series resistors, R1 and R2, are unlike. However, for the symmetrical T, the two series resistors identified as R1 are identical.

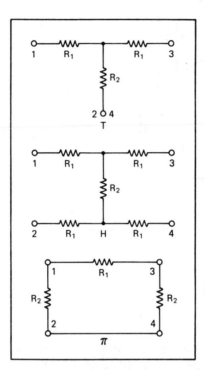

Figure 5-26   T, H, and π pads.

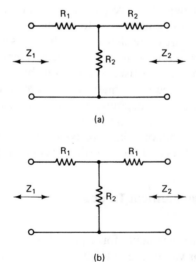

Figure 5-27   (a) Unbalanced T and (b) symmetrical T pads.

TABLE 5-1 DESIGN VALUES FOR PADS

Loss, dB	T PAD		H PAD		π PAD	
	R_1	R_2	R_1	R_2	R_1	R_2
0.1	3.58	50,204	1.79	50,204	7.20	100,500
0.2	6.82	26,280	3.41	26,280	13.70	57,380
0.3	10.32	17,460	5.16	17,460	20.55	34,900
0.4	13.79	13,068	6.90	13,068	27.50	26,100
0.5	17.20	10,464	8.60	10,464	34.40	20,920
0.6	20.9	8,640	10.45	8,640	41.7	17,230
0.7	24.2	7,428	12.1	7,428	48.5	14,880
0.8	27.5	6,540	13.75	6,540	55.05	13,100
0.9	31.02	5,787	15.51	5,787	62.3	11,600
1.0	34.5	5,208	17.25	5,208	68.6	10,440
2.0	68.8	2,582	34.4	2,582	139.4	5,232
3.0	102.7	1,703	51.3	1,703	212.5	3,505
4.0	135.8	1,249	67.9	1,249	287.5	2,651
5.0	168.1	987.6	84.1	987.6	364.5	2,141
6.0	199.3	803.4	99.7	803.4	447.5	1,807
7.0	229.7	685.2	114.8	685.2	537.0	1,569
8.0	258.4	567.6	129.2	567.6	634.2	1,393
9.0	285.8	487.2	142.9	487.2	738.9	1,260
10.0	312.0	421.6	156.0	421.6	854.1	1,154
11.0	336.1	367.4	168.1	367.4	979.8	1,071
12.0	359.1	321.7	179.5	321.7	1,119	1,002
13.0	380.5	282.8	190.3	282.8	1,273	946.1
14.0	400.4	249.4	200.2	249.4	1,443	899.1
15.0	418.8	220.4	209.4	220.4	1,632	859.6
16.0	435.8	195.1	217.9	195.1	1,847	826.0
17.0	451.5	172.9	225.7	172.9	2,083	797.3
18.0	465.8	152.5	232.9	152.5	2,344	772.8
19.0	479.0	136.4	239.5	136.4	2,670	751.7
20.0	490.4	121.2	245.2	121.2	2,970	733.3
22.0	511.7	95.9	255.9	95.9	3,753	703.6
24.0	528.8	76.0	264.4	76.0	4,737	680.8
26.0	542.7	60.3	271.4	60.3	5,985	663.4
28.0	554.1	47.8	277.0	47.8	7,550	649.7
30.0	563.0	37.99	281.6	37.99	9,500	639.2
32.0	570.6	30.16	285.3	30.16	11,930	630.9
34.0	576.5	23.95	288.3	23.95	15,000	624.4
36.0	581.1	18.98	290.6	18.98	18,960	619.3
38.0	585.1	15.11	292.5	15.11	23,820	615.3
40.0	588.1	12.00	294.1	12.00	30,000	612.1

(continued)

TABLE 5-1 DESIGN VALUES FOR PADS (*continued*)

Loss, dB	T PAD R₁	H PAD R₂	π PAD R₁	π PAD R₂
0.1	7.2	50,000	3.6	50,000
0.2	13.8	26,086	6.9	26,086
0.3	21.0	17,143	10.5	17,143
0.4	28.2	12,766	14.1	12,766
0.5	35.4	10,169	17.7	10,169
0.6	43.2	8,333	21.6	8,333
0.7	50.4	7,143	25.2	7,143
0.8	57.6	6,250	28.8	6,250
0.9	65.4	5,504	32.7	5,504
1.0	73.2	4,918	36.6	4,918
2.0	155.4	2,316	77.7	2,316
3.0	247.8	1,452	123.9	1,452
4.0	351.0	1,025	175.5	1,025
5.0	466.8	771.2	233.4	771.2
6.0	597.0	603.0	298.5	603.0
7.0	743.4	484.3	371.7	484.3
8.0	907.2	396.8	453.3	396.8
9.0	1,091	329.9	545.5	329.9
10.0	1,297	277.5	648.5	277.5
11.0	1,529	235.5	764.5	235.5
12.0	1,788	201.3	894	201.3
13.0	2,080	173.1	1,040	173.1
14.0	2,407	149.6	1,204	149.6
15.0	2,773	129.8	1,387	129.8
16.0	3,186	113.0	1,598	113.0
17.0	3,648	98.68	1,824	98.68
18.0	4,166	86.4	2,083	86.4
19.0	4,748	75.8	2,374	75.8
20.0	5,400	66.66	2,700	66.66
22.0	6,954	51.72	3,477	51.72
24.0	8,910	40.4	4,455	40.4
26.0	11,370	31.66	5,685	31.66
28.0	14,472	24.87	7,236	24.87
30.0	18,372	19.58	9,186	19.58
32.0	23,286	15.46	11,643	15.46
34.0	29,472	12.21	14,736	12.21
36.0	37,260	9.66	18,630	9.66
38.0	47,058	7.65	23,529	7.65
40.0	59,400	6.06	29,700	6.06

Unbalanced versus Symmetrical π Pads

Figure 5-28 shows a pair of π pads, with the unbalanced unit in part (a) and the symmetrical unit in (b). Unlike the T pads, which use a pair of series resistors, the π pads use a single series and a pair of shunt resistors. Note that the shunt resistors in the symmetrical pad have identical values.

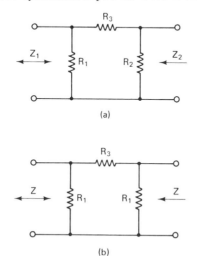

(a)

(b)

Figure 5-28   (a) Unbalanced π pad. (b) Symmetrical π pad.

Balanced H and Symmetrical H Pads

The two drawings in Figure 5-29 are both H pads. The one in part (a) is a balanced pad; that in (b) is symmetrical. The series arms in (a) are made of unequal value

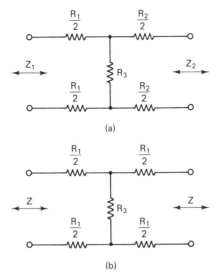

(a)

(b)

Figure 5-29   H pads: (a) balanced and (b) symmetrical.

resistors; the same arms in (b) are identical resistors. Circuitwise, the pads are identical.

Pads can be identified in various ways: by the arrangement of their components, such as in the form of an L, T, or H; by whether they are balanced or unbalanced; by whether they are symmetrical or asymmetrical; by whether they are grounded or not; and by whether they have matching input and output impedances or not. Some of these terms, however, are synonymous.

ATTENUATORS

All attenuators belong to the resistor family and consist of one or more fixed or variable resistors. They are not frequency-sensitive units and are used to control DC or AC voltages or audio signal voltages, doing so by introducing a voltage drop due to current flowing through them and by working as a voltage divider.

The simplest type of attenuator can be a combination of one or more fixed and variable resistors. The name given to specific attenuators often depends on their function, as, for example, a volume control. An attenuator inserted between a sound source and a preamplifier is sometimes referred to as a *loss pad* or sometimes simply a pad. Its function is to introduce a signal loss to avoid overloading the input of the preamplifier. Attenuators are also used on recorder/mixer controls in recording studios and in this function are known as *faders*. Several faders are used, depending on how many channels are available. Ideally, the faders are arranged so that the setting of one does not interfere with the setting of the others. In some recorder/mixer consoles, the mixer circuitry may be used to supply fader action, but the primary function of the mixer, as its name implies, is to mix two or more audio signals.

Attenuators are also impedance-matching devices, with an input impedance matching that of the source and an output impedance matching that of the load.

Insertion Loss

The loss produced by the insertion of an attenuator between a source voltage and its load is referred to as insertion loss, and in the case of a signal, it is measured in decibels. The problem with attenuators is that they are made of resistive elements and so are not amplitude-selective. As a result, all input signals will be weakened, whether those signals are weak or strong.

Filters versus Attenuators

Both filters and attenuators achieve controlled reduction of signal strength. Since the filter uses elements such as coils and capacitors, it takes advantage of the way in which inductive and capacitive reactances function and hence takes advantage of frequency to achieve signal reduction. An attenuator has no reactive elements and uses only resistors; these are not frequency-dependent components.

The T-Type Attenuator

The *T-type attenuator* gets its name from the fact that the combination of its series and shunt arms resembles the letter T, as shown in Figure 5-30; however, because of its configuration, it is sometimes known as an unbalanced attenuator. The unit consists of two variable series arms and one variable shunt arm. As indicated by the dashed lines, all three resistors are variable, are individual units, and are operated by a common shaft. The T-type attenuator can be used with inputs and outputs having the same or unequal impedances.

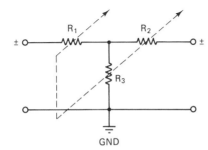

Figure 5-30 Unbalanced T-type attenuator.

The formula for calculating the values of the variable resistors for a T attenuator having identical input and output impedances is

$$R_1, R_2 = \left(\frac{K - 1}{K + 1}\right)Z$$

$$R_3 = \left(\frac{K}{K^2 - 1}\right)2Z$$

where R_1 and R_2 are the ohmic values of the series resistors and R_3 is the value of the shunt arm. Z is the identical input and output impedances in ohms. The value of K is obtained in the same manner as that described for the L attenuator.

The Bridged-T Attenuator

The *bridged-T attenuator*, shown in Figure 5-31, is a variation of the T-type. It consists of two variable resistors, R_5 and R_6, mounted on a common shaft and a pair of fixed resistors, both identified as R_1. The sum of the ohmic values of the series resistors, R_1, is equal to the input impedance. The resistors R_1 should have the same resistance. If the input impedance is 600 Ω, then each resistor R_1 should be 300 Ω. This attenuator is intended to link circuits having identical impedances, and so the input and output impedances have the same value.

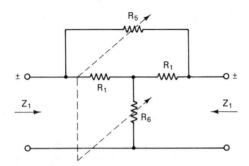

Figure 5-31 Bridged-T attenuator.

R_5 and R_6 function inversely; consequently, as the value of R_5 increases, that of R_6 decreases.

The equation for calculating the component values of the bridged-T attenuator is

$$R_1 = Z$$

$$R_5 = (K - 1)Z$$

$$R_6 = \left(\frac{1}{K - 1}\right)Z$$

where R_1 represents equal-value line resistors whose sum is that of the line impedance; R_5 is the bridging shunt resistor; and R_6 is a shunting variable resistor. The value of K is obtained by the method described in connection with the L attenuator.

The H-Type Attenuator

The H-type attenuator is also known as a balanced unit, since its lines are identical. The upper and lower series arms use resistors of identical value. The shunt arm is tapped at its exact electrical center and is then grounded. Technically, the unit is a pad rather than an attenuator, since all of its elements have fixed values.

The L Attenuator

The *L attenuator*, so named because of its resemblance to the uppercase letter L, is shown in Figure 5-32. The two variable resistors, R_1 and R_2, are ganged, as indicated by the dashed line, and so their shafts are rotated simultaneously.

Regardless of the setting of the attenuator controls, the resistance presented to the source voltage remains constant. At the same time the attenuator drops the source EMF to the amount required by the load. The formula can be stated as

$$R \text{ presented to the source} = R_1 + \frac{R_2 \times R_{\text{load}}}{R_2 + R_{\text{load}}}$$

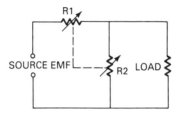

Figure 5-32   L-attenuator.

where R is the total resistance of the attenuator and the load presented to the source voltage, and R_1 and R_2 are the values of variable resistors of the attenuator at a specific setting of the control. R_1 is the series arm; R_2 is the shunt arm.

The L-type attenuator is suitable where both impedances, source and load, are identical. If they are not, then this attenuator can be designed to match either the input or the output impedance but not both at the same time.

If the attenuator is used between two impedances of unequal value, with the input impedance (Z_1) as the larger, the impedance of the attenuator will more closely approximate that of the input. Under these circumstances the values of R_1 and R_2 can be obtained from

$$R_1 = \left(\frac{Z_1}{S}\right)\left(\frac{KS - 1}{K}\right)$$

$$R_2 = \left(\frac{Z_1}{S}\right)\left(\frac{1}{K - S}\right)$$

where

$$S = \sqrt{\frac{Z_1}{Z_2}}$$

K, also known as the K factor, is the ratio of voltage, or power, in decibels, corresponding to a given value of attenuation.

Constant-Impedance Attenuators in Parallel

Attenuators can be combined in parallel, as in the circuit arrangement of Figure 5-33. This diagram shows three T-type attenuators wired in shunt. The line (the input impedance) and the load (the output impedance) have identical resistance values. Table 5-2 supplies the values of R_1 in ohms. Each of the T attenuators is known as a channel, and so the diagram is that of three channels. A greater or lesser number of channels can be used. The table gives data for a minimum number of channels (2) to as many as 6. The insertion loss increases for a larger number of channels. The value of R1 is supplied by

$$R_1 = Z_L\left(\frac{N - 1}{N + 1}\right) \qquad \begin{array}{l} \text{Insertion loss} \\ \text{in dB} = 20 \log_{10} N \end{array}$$

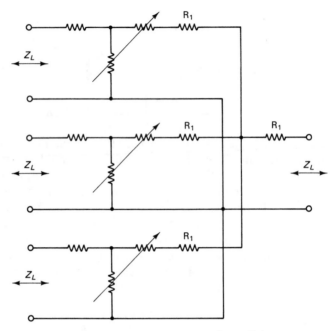

Figure 5-33 T-type attenuators in parallel.

where Z_L represents identical line and load impedances and N is the number of channels in parallel.

TABLE 5-2 VALUES IN OHMS

Z	Number of Channels				
	2	3	4	5	6
30	10	15	18	20	21.5
50	16.6	25	30	33.3	35.7
150	50	75	90	100	107
200	66.6	100	120	133	143
250	83.3	125	150	166	179
500	166	250	300	333	357
600	200	300	360	400	428
Network dB Loss	6	9.5	12	14	15.5

6

Analog and Digital Sound Transformations

In nearly all phases of audio technology, professional and consumer alike, traditional analog (continuous-signal) recording, processing, and signal-transmission methods are rapidly being augmented by digital (computer-like) pulse signals.

A notion is prevalent that digital is about to replace analog. This is not true. Analog and digital methods are used in a joint effort, but although the total contribution of each is not equal, the fact remains that the end product of audio processing must be analog because we hear sounds in analog, not digital. Our ears and minds make sense out of analog signals. The music recording and reproduction chain starts and ends with analog signals. The intervening steps can be digital.

Following the beginning of the audio process, the audio signal is changed to digital by an A/D (analog-to-digital) converter. At some stage before the sound is delivered to headphones or speakers, the reverse process takes place. The sound is converted from digital to analog by a D/A converter.

The advantage of the additional sound processing is that the digital signal is relatively impervious to the ills that have always degraded audio signals. Such problems as limited dynamic range, noise, distortion, and wow and flutter incurred while processing, recording, mixing down, mastering, playing back, and amplifying—virtually every stage in the process of producing and storing music—have been solved through the use of some form of digital electronics. In the transition from phono records to compact discs, record wear, needle scratch, wow, and flutter have been eliminated. Dynamic range, the sonic distance between the softest and loudest sounds, has been extended from 50 to 60 dB to as much as 85 to 90 dB. It is difficult to predict the expected working life of a compact disc, if it receives reasonable care.

The microphone, the start of the whole audio process, is still immune to the digital process. Its input is analog, and its output is also analog.

The Meaning of Analog

The word *analog* (or analogue) means "something that is similar to something else." Producing a graph of an audio voltage is the pictorial representation, or analog of that voltage. The sound from a loudspeaker is the analog of the voltage producing it. The odometer in an auto supplies data about the speed of the car. The numbers indicated by the odometer are analogous to the rate of rotation of the wheels. Analog data are continuously variable.

The Meaning of Digital

The word *digital* has to do with discrete numbers. The instantaneous value of a voltage is digital, since it is expressed in numerical form. The reference is zero, and the instantaneous value is some amount measured at a specific moment in time.

THE BINARY SYSTEM

The most commonly used digital system is the decimal system. Purchases are made, bills are paid, salaries are earned, and taxes are determined, all using the decimal system, a system that contains only 10 digits, 0 through 9. All other digits in this system are combinations of these numbers.

There are various other number systems, the trinary, quaternary, octonary, and so on. The trinary number system uses three digits: 0, 1 and 2. The quaternary uses four digits, 0, 1, 2 and 3. The systems used in electronics are decimal and the binary comprising only two digits, 0 and 1.

The numbers 0 and 1 are referred to as binary digits. The word *bit* is a contraction of binary digit, using the first two letters of the first word and the last letter of the second word.

Nibbles, Bytes, Half-Words, Words, and Double Words

Reference is seldom made to a single bit. A nibble consists of 4 bits; a typical example is 0110. A byte is made up of 8 bits, such as 01000101. A half-word is 16 bits, a word is 32 bits, and a double word is 64 bits. Whether in a nibble or a double word, the bits follow each other in direct succession, not interrupted by punctuation.

Binary numbers are expressed only in two digits, 0 and 1. Other numbers, such as 2, 3, or 4, do not belong in the binary system, and their presence indicates an error.

Symbols

The ten digits 0 through 9 used in the decimal system are the symbols of that system. The two digits 0 and 1 in the binary system are also known as *symbols*. The symbols in the binary system are also used in the decimal, but the reverse is not true. Only two of the decimal symbols are found in binary. Because of the presence of the symbol 9, a number such as 101019 is a decimal number. To emphasize the fact that these symbols are decimal, the number 101019 is sometimes written as 101019_{10}. The subscript 10 is included only if there is some possibility of confusion. A number such as 10101 could be either binary or decimal, since its symbols appear in both systems. If there is any possibility of confusion, the number can be written as 10101_2 to indicate it is a binary number or as 10101_{10} to emphasize that it is a decimal number. The number 2 or the number 10 used as a subscript is known as a *radix*.

Radix, or Base

The radix, or base, of a number system is the number of symbols it uses. The radix of the decimal system is 10 because it uses 10 symbols. The binary number system has a radix of 2. The radix is not involved in any arithmetic process on the numbers.

DECIMAL VALUES OF BINARY NUMBERS

Consider a binary number such as 111. The first symbol on the right has a decimal value of 2^0, the number in the center has a value of 2^1, and the symbol at the left has a value of 2^2. Table 6-1 lists powers of 2 and their decimal values.

TABLE 6-1 POWERS OF 2

$2^0 = 1$	$2^5 = 32$
$2^1 = 2$	$2^6 = 64$
$2^2 = 4$	$2^7 = 128$
$2^3 = 8$	$2^8 = 256$
$2^4 = 16$	$2^9 = 512$

TABLE 6-2 POSITIONAL POWERS OF 2 AND THEIR DECIMAL EQUIVALENTS

2^{10}	2^9	2^8	2^7	2^6	2^5	2^4	2^3	2^2	2^1	2^0	Powers of 2
1024	512	256	128	64	32	16	8	4	2	1	Decimal Value

Since binary 1 indicates the presence of some power of 2, binary 0 is the absence of any such power. Binary 1101, for example, is:

$$1 \quad 1 \quad 0 \quad 1$$

$$2^3 + 2^2 \quad\quad + 2^0$$

$$= 8 + 4 + 0 + 1$$

$$= \text{decimal } 13$$

Table 6-2 shows the decimal equivalent of each place in binary 11111111111. The decimal value of each digit depends on its horizontal position. Digit 1 at the

TABLE 6-3 BINARY PLACE VALUES AND EQUIVALENTS OF DECIMAL NUMBERS

	Place Value				
Decimal Value	16	8	4	2	1
0	0	0	0	0	0
1	0	0	0	0	1
2	0	0	0	1	0
3	0	0	0	1	1
4	0	0	1	0	0
5	0	0	1	0	1
6	0	0	1	1	0
7	0	0	1	1	1
8	0	1	0	0	0
9	0	1	0	0	1
10	0	1	0	1	0
11	0	1	0	1	1
12	0	1	1	0	0
13	0	1	1	0	1
14	0	1	1	1	0
15	0	1	1	1	1
16	1	0	0	0	0
17	1	0	0	0	1
18	1	0	0	1	0
19	1	0	0	1	1
20	1	0	1	0	0

TABLE 6-4 BINARY NUMBERS AND DECIMAL EQUIVALENTS

Binary	Decimal
0000	0
0001	1
0010	2
0011	3
0100	4
0101	5
0110	6
0111	7
1000	8
1001	9
1010	10
1011	11
1100	12
1101	13
1110	14
1111	15

extreme right has a decimal value of 1. Digit 1 at the extreme left has a decimal value of 1024. Thus, the decimal value of 11111111111 is 1024 + 512 + 256 + 128 + 64 + 32 + 16 + 8 + 4 + 2 + 1 = 2047. A power of 2 exists only if bit 1 appears in a column. The particular power of 2 does not exist if it is represented by 0 in that column.

As another example, binary 0010000101 = 133.

$$0 \quad 0 \quad 1 \quad 0 \quad 0 \quad 0 \quad 0 \quad 1 \quad 0 \quad 1$$
$$2^7 \qquad\qquad\qquad 2^2 \quad\; 2^0$$

$$2^7 = 128$$

$$2^2 = 4$$

$$2^0 = 1$$

$$128 + 4 + 1 = 133 \text{ (decimal)}$$

Note that the zeros to the left of bit 1 in the 2^7 position have no effect on the final answer. 0010 has the same binary value as 10. Place values of binary numbers are listed in Table 6-3.

As another example of binary-to-decimal conversion, consider the binary number 10001111000. The decimal value is as follows:

$$1 \quad 0 \quad 0 \quad 0 \quad 1 \quad 1 \quad 1 \quad 1 \quad 0 \quad 0 \quad 0$$
$$2^{10} \qquad\qquad\; 2^6 \;\; 2^5 \;\; 2^4 \;\; 2^3$$

$$2^{10} = 2 \times 2 \times 2 \times 2 \times 2 \times 2 \times 2 \times 2 \times 2 \times 2 = 1024$$

$$2^6 = 2 \times 2 \times 2 \times 2 \times 2 \times 2 = 64$$

$$2^5 = 2 \times 2 \times 2 \times 2 \times 2 = 32$$

$$2^4 = 2 \times 2 \times 2 \times 2 = 16$$

$$2^3 = 2 \times 2 \times 2 = 8$$

$$1024 + 64 + 32 + 16 + 8 = 1144 \text{ (decimal)}$$

Table 6-4 is a partial listing of binary values and their decimal equivalents. Note that 4 binary bits can be used to represent any decimal number from 0 to 15 inclusive, a total of 16 numbers. Both 0 and 1 are bits and are correctly referred to as the *0 bit* and the *1 bit*. The 0 bit is sometimes called the *no bit*; the 1 bit is spoken of as *a bit*.

The ability to change from decimal to binary is a first step in converting an audio voltage waveform, ordinarily expressed as a succession of decimal values, to binary. The waveform is analog and could represent the output voltage of a mic or an audio amplifier.

DIGITAL REPRESENTATION OF A WAVE

The drawing in Figure 6-1 illustrates a single complete sine wave. The horizontal line indicates a few of the points where instantaneous values of the wave may be measured. These instantaneous values are vertical lines, and the length of any of these lines indicates the amount of instantaneous voltage. These voltages, for example, could be 0, 6, 9, 12, 9, 6, 0 for the upper half of the wave and a similar succession of numbers for the lower half. The symbols indicate that all these voltage values are expressed in decimal. These numbers can be converted to their binary form as follows:

$$0 \text{ (decimal)} = \quad 0 \text{ (binary)}$$

$$6 \text{ (decimal)} = 0110 \text{ (binary)}$$

$$9 \text{ (decimal)} = 1001 \text{ (binary)}$$

$$12 \text{ (decimal)} = 1100 \text{ (binary)}$$

Each of the instantaneous values can be represented decimally by 0, 6, 9, and 12 for the first half of the wave and also for the second half. Alternatively, each of the instantaneous values can be represented by 0000, 0110, 1001, and 1100.

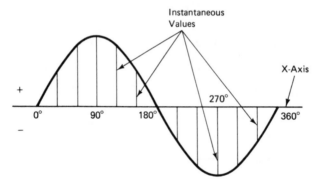

Figure 6-1 Instantaneous values of a sine wave.

The Difference Between Analog and Digital Signals

The conversion of instantaneous analog voltages to corresponding binary form is just the first step in moving from analog to digital form. The binary symbol 0 represents the absence of a voltage; the symbol 1 represents its presence. This can also be shown by a waveform, as in Figure 6-2. This waveform is actually a series of pulses, all of which have the same amplitude. The symbol 0 is indicated by a short, horizontal line, with each 0 having the same length. The symbol 1 is represented by a rectangular pulse, with each symbol 1 the same as any other.

Figure 6-3 shows the pulse representation of binary numbers. The first is 010, corresponding to decimal 2. The next is 001, corresponding to decimal 1.

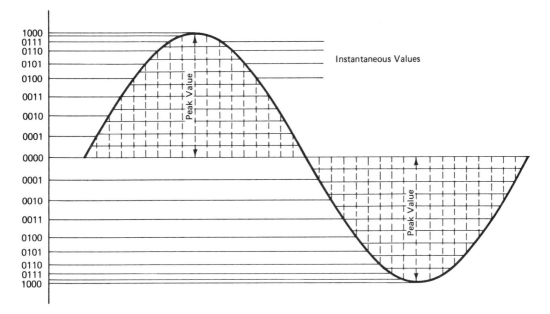

Figure 6-2   Quantization of a sine wave using a 4-bit code.

0 1 0   0 0 1   0 1 1   1 1 0

Figure 6-3   Pulse representation of binary numbers.

The third is 011 (equivalent to decimal 3), and the last, 110, is equivalent to decimal 6. Each of these could correspond to the instantaneous decimal value of a voltage wave.

Analog and Digital Waveforms

Figure 6-4(a) shows a representative analog waveform. By selecting a number of instantaneous values and converting them to their binary equivalents, a new waveform can be produced, as in part (b). These are two representations of the same series of instantaneous voltages.

There are differences. An analog audio signal is not only continuous; it also varies in amplitude. The digital audio signal consists of a series of rectangularly shaped pulses of constant amplitude. Digital is represented by encoding the analog waveform as a series of binary symbols, 0 and 1.

Both waveforms contain information. The information in one of these formats, such as analog, can be converted into the other using A/D conversion.

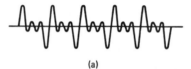

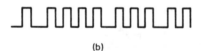

Figure 6-4 (a) Analog waveform and (b) a portion of its binary pulse equivalent.

(a)

(b)

DIGITAL ELECTRONICS FOR COMPACT DISC PLAYERS

Quantization

The conversion of a number of instantaneous voltage values of an analog wave-form into binary digits is called *quantization*. Although the illustration in Figure 6-2 is that of a sine wave, the wave that is quantized—the one whose peak values are constantly measured—is a much more complex type. Because the waveform keeps changing, the question arises as to how often the instantaneous voltage values should be measured.

Sampling

When the quantization of a wave is done repeatedly, the process is known as *sampling*. Discrete-time sampling is done at precisely spaced intervals, and it is essential for the sampling frequency to remain constant. Consequently, the circuit that does this work has its frequency crystal-controlled.

The sampling frequency is 44.3 kHz but multiples of this frequency, such as 44.3 × 2 = 88.6 kHz and 44.3 × 4 = 177.2 kHz, are also used. Multiples of the basic sampling frequency are obtained through the use of a frequency mul-tiplier, a doubler for twice the sampling frequency and a quadrupler for four times.

The basic sampling frequency is a little more than twice the highest audio frequency of the analog waveform, considered to be 20 kHz. Although 20 kHz has been adopted as the top end of the audio spectrum, the scanning frequency has been set at 44.3 kHz, leaving a guard band of 4.3 kHz between twice the upper limit of the audio frequency and the scanning frequency (44.3 kHz − 40 kHz = 4.3 kHz).

The Binary Waveshape

Binary 1 results in a pulse waveform, whereas binary 0 is the absence of a pulse, but both bits, 1 and 0, are referred to as pulses. A binary 0 has as much significance as a binary 1 in a pulse waveform, and the accidental absence of a binary 0 in a pulse waveform can result in a substantial difference in the total binary value of the waveform.

Each instantaneous voltage value uses 16 binary bits when quantizing an analog waveform. If each of these bits has a binary value of 1, the sum of all these bits will be 1 less than 2 raised to the sixteenth power (2^{16}), which is equivalent to decimal 65,535. This means that with 16-bit sampling, the analog waveform can be considered to be divided into 65,535 parts. No binary fractional parts are used, and if the peak amplitude of an instantaneous value results in a fractional binary, it is truncated to a whole number. Thus, if the decimal amount of an instantaneous value is 47,638.23, the decimal 0.23 is dropped. However, the fractional amount, 0.23 in this example, is an extremely small percentage of the number.

These two factors, a high sampling rate that is more than twice the highest audio frequency (20 kHz) and the use of 16 bits for the binary representation of the successive instantaneous amplitudes of the analog wave, mean that this representation is highly accurate. The result is an extremely high dynamic range, usually shown in spec sheets as 90 dB. This is far superior to the dynamic range claimed for phono records or recording tape.

Aliasing

Aliasing is the development of beat note products due to heterodyning between the sampling frequency and ultrasonic audio harmonics. It is referred to as intermodulation distortion (IM) but is generally not included in the specs of a compact disc player.

The Bit Stream

The quantization of an analog wave results in a long series of digital bits with a space between each group of 16 bits, a space that has been established as 22.7 μs. A series of bits is sometimes referred to as a *bit stream*, a *data stream*, or a *string*.

The Analog Envelope

An analog waveform can be represented by a large number of instantaneous values, with each value representing a voltage at a specific moment in time. If the maximum points of these instantaneous values are connected, the envelope of the waveform is obtained. This envelope can be that of the voice of a vocalist, a musical instrument, an orchestra, or any other combination of sound sources. When fed into suitable audio equipment, the envelope reproduces the sound source, whatever that sound may be.

The sound envelope is analog—a rapid succession of instantaneous audio voltages. The voltage values can be converted into a number of binary digits.

These digits are the arithmetic representation of the envelope and, when inserted into suitable equipment, can also be used to reproduce the original sound.

Encoding and Decoding

The process of converting the envelope from its original analog form to digital is known as *encoding*. But since we hear only in analog form, the encoded signal must be converted to its original analog arrangement, a process known as *decoding*. Digital, then, is an intermediary process. An audio component starts with analog and then, to obtain certain desired electronic properties, converts the analog to digital form. At some time prior to sound reproduction by headphones or a speaker, decoding must take place, with the binary form changed back to analog. Digital, then, is a type of sound processing.

It is possible for a sound source to be in digital form, such as a compact disc. But the compact disc is not an original sound source. It is an encoded device and receives its signal from an analog component, such as a phono record. The compact disc, then, is not an original sound source but is simply an intermediate step in the analog-to-digital, digital-to-analog sound process.

Errors

It is possible for both people and machines to make errors when many large-valued numbers are handled in very short periods of time, and the process of encoding and decoding is no exception.

It is also possible for an incorrect binary value to be derived from a particular analog amount. It is true that the percentage of error may be very small, but it is an error nevertheless. The sound difference may be so small and occupy such a small time segment that it will be passed by even a trained listener or someone who is highly competent musically.

A tape deck is used during part of the digital process, and if the tape being used has *dropouts*, missing magnetic particles on the tape, some binary numbers will be missing.

The assumption is made, during the encoding process, that the analog waveform being encoded is perfect. However, it may be distorted; it may include spurious voltages inflicted on it from other sources; or it may include unwanted hum voltages. But all these—and possibly more—form part of the analog envelope and are included when they are encoded.

Bit Quantity

The instantaneous value at any point along an analog envelope is actually a decimal number. The larger the number, the greater the number of binary bits required to represent it. A decimal number such as 20 can be represented by 5 bits. A number such as 1,144 requires 11 bits, and a number such as 65,535 requires the use of 16 bits.

DIGITAL MODULATION TECHNIQUES

A series of random pulses that do not represent data can be said to be *unmodulated*. If these pulses are modified, either by amplitude, by width, by spacing, or by any other technique, and if these modifications represent information, the pulses are then said to be *modulated*. This modulation is different than that used in AM or FM broadcasting, where a radio-frequency carrier is required. No such carrier is needed for digital pulses.

Random Codes

There are various ways of producing electrical pulses. One of these is a random method, such as turning a light on and off. This action produces pulses of current flow to the lamp, but other than that, these pulses do not carry any information, or data. The pulses are random and have no informational significance.

That is not true of pulses that carry data. A telegraph signal consists of coded pulses that carry information. The pulses used in the manufacture of a compact disc do carry information and hence are modulated. This is not to be confused with modulation systems used in radio and television broadcasting. These systems consist of signals that are amplitude- or frequency-modulated, which is a process by which a high-frequency carrier is used to carry a signal. Audio and TV signals are modulated onto a high-frequency wave, which is simply used for transportation. An audio signal has a very limited traveling range, and even a very loud sound signal is not capable of traveling more than a very short distance.

An encoded digital signal is said to be modulated simply because it carries data. These data are supplied in the form of rectangular pulses, and no high-frequency carrier wave is involved.

Pulse Modulation

Pulse modulation is the transmission of data in the form of rectangular pulses. Transmission here means the transfer of data by wire, by cables, or by magnetic tape, with these devices moving the pulse data from one component to another.

The representation of data by pulses can take a number of forms. The pulses can all have equal amplitudes, with the data conveyed by the presence or absence of pulses, by changes in the amplitudes of the pulses, or by changes in their spacing.

There are different kinds of pulse modulation, including pulse-amplitude modulation (PAM), pulse-duration modulation (PDM), pulse-position modulation

(PPM), differential pulse-code modulation (DPCM), delta modulation (DM), and pulse-code modulation (PCM), also known as linear pulse-code modulation (LPCM). Delta modulation is actually just a variation of DPCM.

Of all these different modulation forms, the one used for compact discs and compact disc players is PCM, selected for its advantages over other types. PCM has virtually eliminated cross talk, a situation in which sound from one of the stereo channels, either left or right, leaks over to the other channel. PCM is characterized by a very wide dynamic range, often more than 90 dB. Wow and flutter, those two plagues of phono records, are practically eliminated, down to the point where it is difficult to measure them. Errors that are produced during the PCM process can be corrected.

Delta Modulation

In PCM the amplitude of each instantaneous value is measured. Delta modulation is a technique in which only the difference in amplitude between one instantaneous value and the next is of interest. The difference between one instantaneous value and the next can be very small and so can be represented by a very small number of bits. In PCM, however, each instantaneous value of the analog waveform can be large, consequently creating a demand for a large number of bits.

The problem with delta modulation is that of the voltage difference between successive instantaneous values. Distortion results when the size of the voltage difference is too small for a good approximation and noise is produced if the difference is too large.

Unlike PCM, with delta modulation the analog signal is sampled at 10 to 50 times the highest audio frequency. With 20 kHz accepted as the upper audio frequency limit, the sampling frequency could be 200 kHz to 1 mHz.

Although delta modulation does have a number of advantages in the elimination of some circuitry required by PCM, it does have a lower dynamic range.

Binary Pulse Bandwidth

The bandwidth of audio signals, assumed to extend from 20 Hz to 20 kHz, is approximately 20 kHz, ignoring the very small range from DC to 20 Hz. The bandwidth of a video channel is much wider, being 6 mHz. The band-pass of a quantized binary pulse signal is equal to the product of the number of bits in the binary and the quantizing rate. Commonly, the quantizing rate is 44.1 kHz. If 14 bits are used to represent an average instantaneous analog value, then the band-pass in this example would be $14 \times 44,100 = 617,400$ Hz, or 617.4 kHz. If, instead, the number of bits used is 16, the bandwidth would increase to $16 \times 44,100$, or 705.6 kHz.

Digital Audio Processor

Some manufacturers of compact disc players favor oversampling, using two or four times the basic sampling rate. If the sampling rate is doubled to 88.2 kHz, the maximum band-pass becomes twice that required for the basic band-pass, or $2 \times 705.6 = 1411.2$ kHz, using a string of 16 bits. For a sampling rate that is four times the basic sampling rate, the band-pass becomes $4 \times 705.6 = 2.822$ MHz. Although not every analog number requires 16 bits for conversion to digital, the band-pass is not one that floats—that is, it must accommodate the largest analog number. The band-pass figures quoted here do not consider the reservation of bits for operating requirements, such as those needed for error corrections.

Following quantization, the resulting binary numbers must be recorded on magnetic tape. Neither an audio cassette nor an open-reel tape can supply the necessary bandwidth capability. However, a videocassette recorder (VCR), with its 4-MHz recording capability, can accommodate a binary string resulting from four times oversampling using a 16-bit string. However, although a VCR could be used, there is a special tape recorder made for this purpose known as a *pulse-modulation tape recorder*.

The analog signal can be supplied in several ways: from the amplifier following a microphone, from a phono record, or from the tape of a tape deck. In all instances the analog audio signal is quantized using oversampling, two times oversampling or four times oversampling, with the resulting binary digits either a 14-bit or a 16-bit string. A special tape deck, known as a *pulse-code-modulation tape recorder* (also known as a *digital audio processor*, or DAP) is used for recording the bit strings.

The DAP is preceded by a pair of low-pass filters, one for left-channel sound and the other for right-channel sound. A low-pass filter, as its name implies, is one that passes all signals below a selected frequency and attenuates all frequencies above it. The purpose of this filter is to eliminate aliasing noise, the beat notes resulting from heterodyning action and ultrasonic audio harmonics. It is possible for the difference frequency—the sampling frequency minus the audio signal harmonics—to come within the audio range. Oversampling, using two or four times the basic sampling frequency, helps minimize aliasing noise. The circuit is identical to that previously shown in Figure 5-14.

String Values

In early digital components, 14 binary bits were used for the quantization of binary numbers. The advantage of the PCM recorder is that it could more easily accommodate the high value supplied by an occasional instantaneous peak. However, more important, the PCM recorder did increase the dynamic range substantially. The decimal equivalent of a 14-bit string is 16,383, but that of a 16-bit string is 65,535, which is four times as great.

The need for a 16-bit string is also due to the fact that the binary digits not only represent the desired sound but are needed for signal control as well. Pulses are required for timing, indexing, and locating the first and last tracks on a compact disc. Control information is in *frames*, with the bits in the frames called *channel bits*. A complete frame consists of 588 channel bits. Channels are used, since it is essential for the compact disc player to be able to differentiate between sync bits, parity bits, and audio information.

SAMPLE-AND-HOLD CIRCUITS

Following the low-pass filters, the left- and right-channel signals are fed into a pair of sample-and-hold (S/H) circuits. The sampling rate is either at the basic rate of 44.1 kHz or multiples of two or four times this rate. The function of the S/H circuitry is to act as a delay device to make sure that the left- and right-channel signals remain in step—that is, in phase. The samples of data that are input to the S/H circuitry are at 22.727-μs intervals at the sampling frequency ($f = 1/44.1$ kHz).

ANALOG-TO-DIGITAL CONVERTER

Up to this point the signal has remained in analog form. As shown in Figure 6-5, the next step in the analog-to-binary process is to supply an A/D converter, with the analog voltage values obtained during the quantization process. In this circuitry the analog values are converted into strings of binary digits consisting of 0s and 1s. Thus, the analog signal becomes a 16-bit PCM signal.

At the output of the A/D circuitry there are left and right strings of binary digits corresponding to the analog values obtained during quantization. The digital pulses are not put to use immediately but instead are brought into the memory

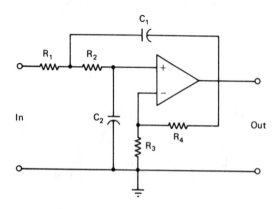

Figure 6-5 Low-pass filter works as an A/D converter and is identical to the circuit previously shown in Figure 5-14.

circuit of a computer that is clock-controlled. The problem with the A/D converter circuitry is that it can produce some errors.

THE CLOCK

A crystal is a mineral that can vibrate at a specific frequency, depending on the dimensions to which it is cut and ground. The selected frequency is highly accurate and can be made to remain extremely constant. Since frequency is a function of time, an oscillator that is crystal-controlled can be made to function as a highly accurate clock.

The waveform produced by a crystal-controlled oscillator is a sine-wave voltage. It can, however, be modified into the form of a sharp pulse, which can be used to trigger the on/off times of various circuits in an accurate manner.

Subcodes

A compact disc player has operational features that must be precisely timed. This is done through the use of subcodes, which are in the form of 8-bit strings. Subcodes contain data pertaining to the functioning of a compact disc and contain no musical information. Including subcodes does not add distortion to the sound information.

Subcodes are used to control the operational information that is displayed on the front of the compact disc player. They are used in the search for wanted bands of music, for the separation of these bands, for repeat playing, and for any other functions that are not musical.

Pulse-Code Modulator

The pulse-code modulator, or PCM, has inputs for a microphone, phono records, or sound supplied by tape decks. For each of these sound sources, the PCM is equipped with buffer amplifiers that not only amplify the input signal but also isolate it from other circuitry. The sound source is selected by a switch located on the front of the PCM. On the same front panel is an input-level control that adjusts the amount of analog signal strength.

Pre-emphasis and De-emphasis

The PCM has a pre-emphasis control, pre-emphasis being a form of signal modification. Pre-emphasis for compact discs is 50-μs treble boost and is used to improve the treble signal-to-noise ratio. In order to obtain a flat frequency response during playback of a compact disc, an equal amount of de-emphasis is supplied in the compact disc player.

Dither

One of the intriguing aspects of playing a compact disc is that its minimum noise level, its noise floor, is so low that it is imperceptible and cannot be heard during the playing of a compact disc. Noise, however, is so prevalent with other forms of electronic sound reproduction that some find its absence disturbing. As a result, some manufacturers deliberately add noise to CD masters, a technique known as *dither*. This is not a totally accepted practice. The sound supplied by the compact disc is practically identical with that recorded on the PCM's tape deck, an advantage negated by dithering.

TIME DELAY

Analog time-delay is being gradually replaced by digital techniques. The advantage is a significant reduction in noise and distortion. The problem with acoustical time delays is that they supply inherent group-delay distortions and low-frequency cutoffs. These were initially replaced by electronic bucket-brigade delay lines.

It was not until the advent of purely digital time-delay techniques using pulse code modulation (PCM) that an S/N of 90 dB could be achieved. This, as well as extremely low distortion, allows digital delay to meet the requirements of studio equipment.

Digital Time-Delay Technique

Time quantization is achieved by sampling the signal periodically. A signal is completely defined by its samples if the sampling rate is at least twice the signal bandwidth. The amplitude quantization results from the conversion of these samples into digital numbers. Rounding off these numbers results in *quantization noise*.

Digital Filtering and Oversampling

The compact disc format specifies that audio be stored on a disc at the rate of 44,100 sixteen-bit words for every second of music, and that is for only one channel. In addition to the musical information, other data must be included, such as EFM (eight to fourteen modulation) synchronization, error-correction codes, and so on. These add to the constant flow of information from the disc. In fact, the optical pickup of every CD player reads data from a disc at the rate of 4.3218 million bits per second.

Ultimately, the digital output must be converted to analog using a digital/analog (D/A) converter. But these converters are not perfect, and so the D/A output is not a precise representation of its digital input. Further, sampled analog signals, such as those that are present at the output of a D/A converter, contain a large amount of ultrasonic information generated by the digital process. This consists of extraneous high-frequency data that are not part of the original audio signal. Further, these data are potentially harmful to high-fidelity components and must be suppressed in some way.

If the sampling frequency is 44.1 kHz in the compact disc player, the signal following demodulation will contain unwanted remnants of the sampling frequency. In addition to the audio frequency band, 20 Hz to 20 kHz, spurious signals will exist, such as 44.1 kHz + 20 kHz; 88.2 kHz + 20 kHz, etc. The result will be that unwanted pulsating noise will exist at regular intervals, beginning at 24.1 kHz (44.1 kHz − 20 kHz), even though the audio spectrum ends at 20 kHz.

Double sampling, also known as oversampling, is a technique in which the sampling rate is increased to 88.2 kHz. By doubling the sampling and passing the digital signal through a special kind of filter, this technique moves the beginning of the first band of ultrasonic noise in the analog signal from 24.1 kHz up to 68.2 kHz. Since this noise no longer begins so soon after the end of the audio frequency band at 20 kHz, it is possible to use a low-pass pass filter having a very gradual roll-off, an inexpensive type having few elements.

A compact disc player may use quadruple sampling. This serves to move spurious ultrasonic frequency components far above the audio band, eliminating possible intermodulation effects (Figure 6-6).

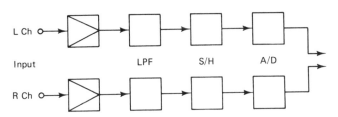

Figure 6-6   Left- and right-channel signal progression: LPF (low-pass filter); S/H (sample and hold); A/D (analog-to-digital conversion).

The Quantization Problem

The CD standard allows for 65,535 different quantization possibilities, but not every sample is precisely represented. The differences between the actual value of a specific sample and its binary equivalent generate quantization noise, which shows itself as a spurious high-frequency signal. In the case of compact discs, quantization noise bands are centered at 88.2 kHz, 132.3 kHz, 176.4 kHz, 220.5 kHz, and so on. Such noise occurs in any digital audio product and is a natural consequence of using sampled waveforms.

It is true that the bands of quantization noise are well above the highest humanly perceptible audio frequency. But they can interfere with an analog tape recording by beating against a bias oscillator or affect stereo radio transmissions. They can also cause instability in the form of oscillation in an amplifier or, in an extraordinary case, burn out the tweeter in a loudspeaker system.

Brickwall Filtering

Early compact disc systems used a *brickwall filter* following the D/A converter in order to eliminate these unwanted frequencies. Although such filters are very effective in removing unwanted noise bands before producing any damage, they had a number of disadvantages. They were bulky, expensive to build, and drifted with age. As a result they had a negative effect on sound quality because of subtle variations of frequency response and not-so-subtle effects on the phase relationships of higher musical frequencies.

To meet this problem, later generations of compact disc players have used a digital filter in addition to the analog filters.

Action of the Digital Filter

A digital filter takes the stream of digital words fed to it and, working like a computer, calculates additional digital words, which it inserts between the original ones from the disc. The filter does this through an elementary series of high-speed multiplications and additions.

This high-speed creation of new digital words is called *oversampling* because the filter creates additional digital words, or samples, where none existed before (Figure 6-7).

A digital filter, then, processes audio data while still in digital form before it is reconstructed as an analog waveform. The major advantage is that the digital filter's computation shifts the unwanted noise bands higher in frequency. The faster the filter's operating speed, the further away the noise bands will be, and progressively less radical analog filters will be required to protect the rest of the

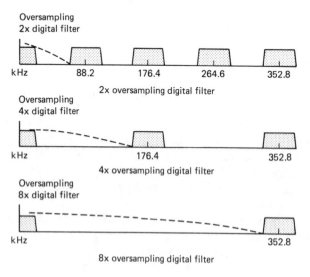

Figure 6-7   Oversampling digital filters (Courtesy Onkyo USA Corporation).

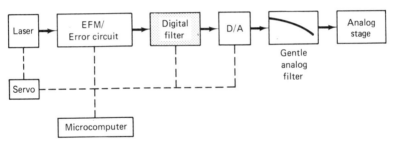

Figure 6-8 Positioning of digital filter in compact disc player (Courtesy Onkyo USA Corporation).

high-fidelity system from quantization noise bands. Thus, the burden of filtration shifts from the analog domain to the digital domain, where it can be done more easily, more accurately, and with little or none of the sonic problems associated with steep slope analog filters. The digital filter is positioned prior to the D/A (Figure 6-8).

Linear 18-Bit Conversion

A 16-bit D/A converter cannot reproduce perfectly the 16-bit audio data from a compact disc. Inevitable conversion losses always result in something less than 16 bits of information returning to the analog domain. On the other hand, a linear 18-bit converter can provide a much more accurate reproduction of those 16 bits, providing measurably lower noise and distortion in the audio signal.

Twenty-Bit Conversion

If 18-bit conversion offers advantages over 16-bit, does 20-bit conversion offer similar advantages over 18? The answer, surprisingly, is no.

The reason for this is that the compact disc carries information in the form of 16-bit words, and the job of any D/A converter is to convert that digital information as accurately as possible. Eighteen-bit conversion addresses the real-world limitations of 16-bit converters, but 20-bit conversion provides no demonstrable benefits.

D/A Conversion Linearity

D/A converter accuracy is even more important than the number of bits processed by the D/A converter. Conversion accuracy, also known as linearity, is the factor that determines the fidelity of the audio signal.

A D/A linearity test measures the converter's ability to reproduce various analog signals, particularly less intense ones such as soft musical passages that directly reflect the value of the digital word from the disc. Any deviation or nonlinearity results in harmonic distortion in the audio signal.

Figure 6-9 Most significant bit (MSB) and least significant bit (LSB) in 16-bit word.

For example, when a bit changes from 1 to 0, the analog output must decrease exactly by a proportional amount. The amount that the analog voltage changes depends on which bit has changed, because each bit has a different influence on the analog signal. The most significant bit (MSB), as shown in Figure 6-9, accounts for a change in fully half of the analog signal's amplitude, whereas the least significant bit (LSB) in an 18-bit word represents a very minor amplitude change of less than four parts in a million.

One-Bit Techniques

MASH (multistage noise shaping) and bit streaming are methods for converting sampled data to an analog waveform using only 1 bit. MASH uses a pulse-width modulation (PWM) signal, whereas bit-streaming technology uses a related pulse density modulation (PDM) technique.

The claim for a technique called third-order MASH is that it can eliminate quantization noise in the audible range. Third-order MASH is, basically, a sophisticated feedback system, with the term third-order indicating that the process occurs three times. The result of third-order MASH noise shaping is to move the quantization noise so high in the frequency spectrum that almost all noise artifacts are removed from the audible range.

Single-bit technology permits a single bit to do the work of many and, since only 1 bit is used, avoids alignment problems. Accuracy is achieved by very fast operation. However, the processor itself introduces digital or quantization noise.

A disadvantage of single-bit conversion is that it is inherently dirtier than multibit designs in that the quantization noise bands are potentially more intrusive. A technique called *noise shaping* uses highly sophisticated computations (algorithms) to alter the noise spectrum of the signal. However, algorithms are a poorly understood branch of applied mathematics. Without proper bit shaping, 1-bit methods are inferior to parallel methods. As of the present time, the greatest advantage of the 1-bit designs is that they are cheaper to produce than multibit methods.

DIGITAL REVERBERATION

Four-channel sound is the most serious attempt to supply listening realism in the home; in prior years it was unsuccessful, not due to technical failures but to the competition between three systems: discrete, QS, and SQ. Various methods, including mechanical springs, were tried, and electronically, a so-called bucket-brigade method was touted. The bucket brigade was found suitable only for short delays of up to 30 ms or so. Cascading bucket brigades to produce delays of hundreds of milliseconds, as required to recreate large-hall acoustics, resulted in excessive degradation in audio quality.

Digital reverberation (reverb) is a more desirable technique and, although it may be tarred with the bucket-brigade brush, is worthy of consideration, since it does overcome the failure of other systems.

Digital reverb can be added to the playback of recorded sound sources in the home via a technique that can be used with the addition of a pair of rear speakers. The system is adaptable to any stereo setup. No additional amplifiers are required, since the reverb system can be accompanied by a built-in pair of integrated units.

In a live performance, reverberant sound arrives at the ears of the listener only a few hundredths of a second later than the direct (dry) sounds but are not perceived separately. They are responsible, however, for the listener becoming aware of the spaciousness of the listening environment. The blending pattern of the direct and reverberant sounds varies both with the location of the listener and the size and reflectivity of the room, the auditorium, or the concert hall.

The digital time-delay system duplicates, electronically, the otherwise absent delayed, reflected sounds. While the program material is heard as usual through the front speakers, the digital reverb unit copies these sounds, delays them, and cross-couples them to recreate the multidimensional paths of sound, supplying a reverberant field of high-echo density. The length of delay and the amount of sound recirculation is under the control of the listener.

Digital time delay works by capturing brief samples of the dry sound, storing it for a small fraction of a second, and then releasing it. In the unit itself, the audio signals are encoded into digital bits and are then fed into a random access memory (RAM) device. The digital bits remain stationary within the RAM until they are moved to a second RAM, and so on through the series. At three different stages they are retrieved and decoded to recover the audio signal. This process happens thousands of times each second. The amount of time delay is determined by the rate at which the data are fed into the RAMs.

Although this in itself is not a new idea, time-delay units are available with memory-equipped delta modulation. Instead of generating coded groups of pulses at regular intervals to represent the amplitude of the audio signal voltage at each moment, delta modulation uses a waveform detector to digitally encode the moment-to-moment changes in the audio signal voltage. This method requires the use of fewer pulses.

7

Acoustics

THE MEANING OF ACOUSTICS

Acoustics is a combination of arts and science whose subject matter is the production, control, transmission, reception, and utilization of sound energy. It also includes the transduction of sound from its original form as sound energy into some other form, notably electrical energy, plus the converse of this energy transformation—that is, electrical energy transformed into an equivalent sound energy. Acoustics also includes a study of the effects of enclosures and open spaces on sound and the positioning and use of various types of microphones and loudspeakers. In a broader sense, acoustics also includes a study of music, musical instruments used singly or in combination, and the production of different kinds of sound. Acoustics can include the generation of new kinds of sound, the minimization of noise, or the deliberate introduction of noise as a music form. Acoustics can include a study of electronics in general and, more specifically, the development and utilization of musical synthesizers. The art and science of acoustics, then, would seem to be practically limitless.

Composite Sound

Dry sound, also called primary sound, is that heard directly from a sound source. It would seem that persons positioned close to a noise source would hear only dry sound, but even in this situation there is some ambient noise. Depending on the location of the noise source and its amplitude, the dry sound is more or less intelligible. Dry sound is more prevalent in the front rows of an auditorium or all the seats in an open-air amphitheater. It is also the main sound source for those sitting very close to loudspeakers or those wearing headphones, assuming that it is only dry sound that has been recorded or transmitted.

More commonly, sound consists of a combination of dry and reverberant sound (wet sound). Reverberant sound consists of sound reflected from a variety of barriers. Not all sound is reflected, depending on the absorption characteristics of the sound-reflecting materials. Alternatively, some sound consists of a number of reflections before it is heard.

Composite sound consists of the sum of the sounds reaching the listeners. These sounds can be partially or completely in phase, both states depending on the arrival times of the sounds at the ear of the listener. Composite sounds can be wholly or partially out of phase, in which case the sounds are more or less subtractive. The total resultant sound, when in or out of phase, depends not only on the phase relationships of the dry and wet sound but also on the amplitude of each at the moment the sounds are joined. Where the two sounds are 180° out of phase and have equal instantaneous amplitudes, the resultant sound is zero.

REVERBERATION TIME

A reflective area may be smooth, pebbled, or wrinkled. The structure may also be concave, in which case it tends to focus the reflective sound. If concave, the diameter of the reflector must be greater than the wavelength of the incident sound.

Because both direct and reflected sound reach a microphone, sound coloration takes place right at the start. Since this is the sound that will be recorded, what listeners will ultimately hear at home will be this composite sound, to which will be added some reflective sound produced in the listening room. This is not the sound heard by a listener in an auditorium.

One of the conditions for the existence of reverberant sound is an enclosed space. Unless surrounded by objects capable of reflection, outdoor sound is mostly nonreverberant. In an enclosed space, sound waves are repeatedly reflected by boundary planes—that is, walls, the floor and ceiling, and any other reflective surfaces.

After a short time, a complete intermixing of sound waves, which join each other in all directions, takes place. At a sufficient distance from the sound source, the space between it and the reflective surfaces is (presumably) uniformly filled with sound. This uniformly sound-filled space does not occur in the immediate proximity of the sound source. Here the direct sound transmitted by the source prevails over reflected sounds. A mic placed in proximity to the source reacts more to the direct sound and less to reflected waves. To change the ratio between direct sound and reverberation, the distance of the mic to the source has to be varied; in many instances, especially in large enclosures, this isn't practical.

Optimum reverberation time varies for particular types of music. In general, reverberation times not exceeding 1.5 s are optimum for the recording of classical and modern music. Rooms for pop or jazz should have a reverberation time below 0.8 s for recording purposes (Table 7-1).

TABLE 7-1   TIME REQUIRED FOR A SOUND WAVE TO TRAVEL TO A REFLECTING SURFACE AND BACK TO LISTENERS AT VARIOUS DISTANCES

Total Distance Traveled (feet)	Distance from Ear to Barrier (feet)	Approximate Time Required (seconds)
11.2	5.6	0.01
22.4	11.2	0.02
33.6	16.8	0.03
44.8	22.4	0.04
56.0	28.0	0.05
63.2	31.6	0.06
78.4	39.2	0.07
89.6	44.8	0.08
100.8	50.4	0.09
112.0	56.0	0.1
224.0	112.0	0.2
336.0	168.0	0.3
448.0	224.0	0.4
560.0	280.0	0.5
632.0	316.0	0.6
784.0	392.0	0.7
896.0	448.0	0.8
1008.0	504.0	0.9
1120.0	560.0	1.0

Paths of Direct and Reverberant Sound

As indicated in Figure 7-1(a), the paths taken by direct and reflected sounds can become quite complex. Reflected sound can move directly to the ears of the listeners, but sometimes there are multiple reflections, with the sound bouncing from one surface to another.

Every time sound strikes a reflecting surface, it loses some of its energy. It also loses some energy in its travels from the reflecting surface to the listener. Reflected sounds arrive only a few hundredths of a second later than direct sounds and are not perceived separately by listeners. However, they do create the effect

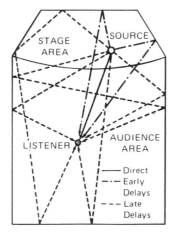

Figure 7-1 (a) Some possible paths of direct and reverberant sound (Courtesy Gould Inc., Audio Pulse Division).

of listening in an enclosed environment. The blending pattern of these sounds, direct and delayed, varies with the location of the listener and the size and reflectivity of the listening area.

CONTROLLED REVERBERATION

Reverberation can be controlled directly in a concert hall or listening room in the home through physical modification of the listening environment or through devices that add artificial reverberation. Time delay can also be added directly via electronic devices to an existing in-home hi-fi system. Not only is the richness of the played-back sound augmented, but the reverberation can be adjusted in order to simulate that of a concert hall, a small or large auditorium, a church, or any other enclosure in which music is being produced.

The in-home artificial reverb unit works by capturing a brief sample of the primary sound, storing it for a small fraction of a second, and then releasing it without altering its character or quality.

HAAS, OR PRECEDENCE, EFFECT

When listening to music, it is mainly through direct sound that we are able to determine the positions of the instruments. This is due to the fact that the human brain establishes sound location based on the arrival time of sound. Reflected sound always takes longer to reach our ears than direct sound. After instrumental or vocal direct sound has been located, it can move, but its perceived direction tends to remain the same for some time. This phenomenon is referred to as the *Haas*, or *sound precedence*, *effect*.

Acoustic Treatment

It is easy enough to apply sound absorbent materials to the walls of a room or acoustic tiles to a ceiling. That does not ensure that the resulting sound absorption will be satisfactory. The best arrangement is to have a reverberation time that is a bit higher at the bass and treble ranges and somewhat less at the midrange. The reason for this lies in the frequency selectivity of the human ear for low-level sounds.

RESONANCE

All materials have a natural resonant frequency, a frequency at which they can be most easily made to vibrate. That vibration may be readily visible, as in the case of a violin string, or it may be invisible. If the resonance body is a large wooden panel, the resonance may not be visible, but it may be felt by touching the panel lightly with a finger tip.

Resonance is the vibration of an object caused by the application of an applied stimulus, such as the energy contained in a sound wave. The amplitude of the vibration can be fairly large. The vibration can intensify or enrich the original sound producing it. The mouth can produce a condition of resonance due to its construction, a type of resonance-chamber arrangement of the pharynx aided, in some instances, by the nostrils. Percussion of the chest can also result in resonance. Resonance can occur in a room or hall, thus emphasizing a particular frequency. In a studio or music hall, resonances are caused by reverberant sound that is in phase with incident direct sound.

A number of resonances can occur at different frequencies, depending on the mass, size, and materials of which an object is made. When loudspeakers resonate, they produce the amplified equivalent of one or more frequencies, a condition that is eliminated by supplying additional supports to the inner frame of the speaker enclosure or by packing the interior with materials that are highly sound absorbent. Any enclosure is capable of supporting not only one but a number of resonances, with these at various frequencies. At the fundamental resonant frequency, sound levels can increase as much as 10 to 20 dB.

Antiresonance

The opposite of resonance, antiresonance, can occur when direct and reverberant sound are out of phase—that is, when their phases are separated by 180°. These are referred to as *nodes*, or *nodal points*, whereas sound reinforcement is said to have *antinodes* or to be in *antinodal condition*. Thus a node or nodal point is an area that is free of vibration or that has had the amplitude of its vibration reduced. An antinodal area is one that has augmented sound vibration. The in-phase or out-of-phase condition may be either complete or partial, depending on the number of degrees by which the two sounds are in or out of phase.

Eigentones

Room resonances are also known as standing waves, or eigentones. They can result in hum pickup in a studio if a microphone is positioned at an antinode. Although hum is generally considered as having a frequency of 50 or 60 Hz, the unwanted sound is more likely to be either 100 or 120 Hz.

Not only can walls, floor, and ceiling be resonant, but a condition of room resonance can also exist. Such a condition occurs most often when the walls are parallel and the rooms are bare, having no furniture, wall coverings, rugs, decorative materials, seats, or people occupying those seats (Figure 7-1b).

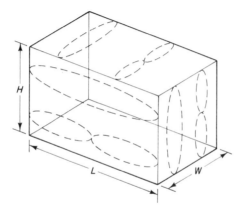

Figure 7-1 (b) Resonant condition when parallel walls are a half-wavelength apart. Drawing shows resulting eigentones.

Forced Resonance

The stimulation of a material into resonance is sometimes referred to as *forced resonance*. At one particular frequency, the material will vibrate strongly, and somewhat less so at higher or lower frequencies. The amount of vibration is identified by the letter Q; a material having a high Q is one that will vibrate strongly at its resonant frequency, also known as its *natural resonant frequency*.

Determination of Room Resonances

The average velocity of sound in the air of a hall or auditorium is useful in the determination of possible room resonances. This is done by dividing the velocity of sound by the dimensions of the room, multiplied by 2 since the resonances take place at one-half wavelength. Assuming a room has different length, width, and height, there will be a minimum of three possible resonant frequencies.

Consider a room measuring 25 ft wide by 12 ft high by 30 ft long. The basic resonant conditions would be:

$$\frac{1117}{2 \times 25} = 22 \text{ Hz} \qquad \frac{1117}{2 \times 12} = 47 \text{ Hz} \qquad \frac{1117}{2 \times 30} = 19 \text{ Hz}$$

where 1117 is the velocity of sound in air if the air temperature is 59°F.

These are the fundamental resonant frequencies. There would also be harmonics of these frequencies. For the second harmonic, multiply the fundamental resonance by 2, for the third harmonic multiply by 3, and so on. The results of this arithmetic would demonstrate that the resonant frequencies are fairly close to each other.

Room resonances can also be calculated with metric measurements. The basic formula is

$$f = \frac{170.25}{L}$$

where f is the frequency in hertz, 170.25 is one-half the velocity of sound in meters (340.5/2), and L is the dimension of any selected room surface, such as a wall, ceiling, or floor, measured in meters.

The resonant frequencies are sometimes described as *natural*, or *eigen, frequencies*. As an example, assume a wall that is 30 ft long. Since 30 ft = 9.144 m, the natural resonant frequency of this wall, in hertz, will be

$$f = \frac{170.25}{9.144} = 18.62$$

This corresponds closely to the data supplied in the preceding problem. Since the space available for a recording studio is not always specifically designed for such use, it may not be easy to make acoustical changes in order to produce substantial changes in reverberation time. Measuring the resonant frequencies in advance of a commitment to use a space can be helpful.

Room Modes

Depending on its construction and geometry, a room used for musical playback may tend to favor certain resonances, a condition known as a *room mode*. The resonances are closely spaced frequency-wise, and if these frequencies are present in the music being played, the room resonances will reinforce them. Room resonances can be eliminated or minimized by altering parallel surfaces.

Efficiency of Energy Transfer

Musical sounds begin with an energy source, which can be mechanical or electrical. The conversion of mechanical or electrical energy to sound energy is highly inefficient, with the efficiency at about 1%—that is, the inefficiency is approximately 99%. Nearly all the energy is expended in heat. Thus a 75-piece orchestra, playing at full strength, produces the equivalent in electrical power of 70 W, or barely enough energy to light a small room. A piano is capable of producing the equivalent of less than $\frac{1}{2}$ W in terms of electrical power.

Since sound energy dissipates rather rapidly upon leaving a source, very little of it reaches the ears of a listener.

ECHO AND REVERBERATION

Echo and reverberation are both aspects of reflected sound and for this reason are sometimes considered synonymous. However, there is an important difference. The effective reverberation of a sound is not separately noticeable or distinct from direct sound. Further, it enriches direct sound, making it appear fuller. The time span between a direct sound and an echo is much larger than that of reverberant sound, and so an echo is noticeable.

Flutter Echo

An *echo* can be a single, distinguishable reverberant sound and is quite evident following the complete attenuation of the original dry sound. A *flutter echo* consists of multiple sound reflections between parallel surfaces, with each echo distinguishable from those that follow. If the flutter echo is periodic, it is sometimes referred to as a musical echo.

Sometimes an echo consists of a single sound sent back toward the sound source. The delay time of a sound for the formation of an echo is about 50 milliseconds or more, and a time delay of a full second or more is not uncommon. The echo is generally a sound reflection from a hard surface. It is possible to hear multiple echoes when the first-formed echo is repeatedly reflected from other surfaces.

The Sound Shadow

Whether a sound will be reflected or not from a surface is dependent on the substance of which the material is made and also on its area compared to the wavelength of the sound. For reflection to take place, a reflective surface must be larger than the wavelength of the incident sound.

If the area of the reflective surface is smaller than the wavelength of the incident sound, the sound will bend around that surface. Since the sound will have a variety of frequencies, the wavelengths will be different. Consequently, some of the sound will be reflected and some will not. Since lower frequencies have greater wavelengths, these will not be reflected. The midrange and treble tones and their harmonics will form part of the reverberant sound. The open region behind the reflecting surface will then be a sound-shadow area.

Amount of Sound Reflection

Assuming a noiseless environment, reverberant sound consists of sounds reflected from a variety of surfaces—walls, ceiling, floor, seats, carpeting, lighting fixtures, pictures, people and their clothing. The amount of sound reflection depends on the softness or hardness of these surfaces, whether they are smooth or rough in texture, their absorption characteristics, the area of their surfaces, and whether those surfaces are fixed in position or are in motion. The human skin is a sound reflector and is better in this respect than cloth.

The Anechoic Chamber

An *anechoic chamber* is a closed room in which only dry or direct sound can exist. The walls are covered with a material that is highly sound absorbent. Audio measurements are made in such a room, with the elimination of reverberant sound as an uncontrollable variable.

The Echo Chamber

An echo chamber is the opposite of an anechoic chamber and is a windowless, closed room whose ceiling, floor, and walls are hard and are covered with materials having maximum sound-reflection qualities.

In-home Acoustics

There are a number of variables that affect the acoustics of a room used in the home either for playback or for recording. These factors include the size of the room, its geometry (its shape, whether square or rectangular, and whether the walls are continuously straight or have curves or rectangular insets), the amount and positioning of the furniture, the number of occupants and whether they are sitting, standing, or moving, the kind and amount of clothing worn by the occupants, rugs (or their absence) and the materials of which the rugs are made, and objects in the room that may have a resonant frequency in the audio range.

Because of this large number of reverberant variables, it is unlikely that any two rooms will have the same acoustics. It also explains why a loudspeaker heard in a dealer's showroom can sound so different when purchased and used in the home.

Importance of Microphone Positioning

The amount of pickup of reverberant sound by a microphone depends on its placement with respect to the reverberant source. A certain amount of reverberant sound is desirable, since, correctly used, it enhances dry sound. However, excessive reverberance can make speech wholly or partially unintelligible.

Greater reverberant pickup by a microphone can be had by increasing the working distance, the separation between the mic and sound source. However, if the microphone is too far from the sound source, the result can be ambient sound overwhelming the direct sound. There may also be an increase in echoes and ambient noise. Unwanted sound is often more noticeable during playback.

Size of Reflective Object and Frequency

Low-frequency waves may impinge on an object whose dimensions are smaller than the wavelength of the sound. In this case the sound is not reflected. For higher-frequency sound waves, reflection takes place. This does not necessarily mean the sound will be reflected to the source. Whether it will or not depends on the angles of incidence and reflection. The sound may be bounced off a number of surfaces until its energy is completely dissipated in the form of heat. The energy is not lost, but it is converted to a different form. If the reflection of the sound and its absorption are complete, there will be a quiet zone, or a so-called dead spot behind the reflecting object, working in this case as a sound barrier.

Hot Spot

A *hot spot* in any enclosure is an area in which direct and reflected sound appear to be focused and in which the phasing of the two is additive. The resultant sound appears to be excessively loud.

Dead Spot

A *dead spot* in any enclosure is an area in which direct and reflected sound are substantially out of phase. It could also be any area in which materials are used that are highly sound absorbent, with a minimum of reflectivity. This combination of phase cancellation plus absorptivity produces an area in which sound is very weak. If the sound is music, it will seem to be very low; if it is speech, it will be practically unintelligible. Sometimes, when moving away from a hot spot or dead spot, there will be a noticeable decrease or increase in sound level (Figure 7-2).

Compared to radio waves, sound waves are fairly long and can be as much as 50 ft for bass tones, decreasing to about 1 in. for the treble range. Reflective surfaces depend not only on the materials of which they are made but on the total surface area.

Behavior of Dry Sound

Dry sound, the sound leaving a source, is affected by the size of that source. If that source is physically smaller than the wavelengths of the sound it produces, the sound will travel outward in all directions with approximately equal amounts of energy. Assuming no barrier or reflecting surface directly behind the source,

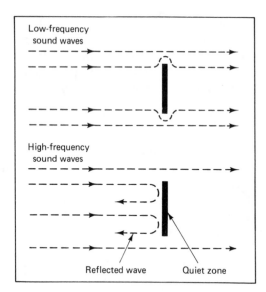

Figure 7-2 Formation of a dead spot.

the sound will be in the shape of a sphere. Since the sphere will be expanding, the energy of the sound over the larger surface area will decrease rapidly. The area of a sphere is proportional to the radius squared, and since the radius is equivalent to the distance traveled by the sound, the loss of energy is proportional to the square of that distance.

However, if the sound source is large compared to the sound wavelength, the outgoing wave is flat rather than spherical. The wavelengths produced by a sound source range from about 12 feet (3.6576 meters) to less than 1 foot (0.3048 meter). Unlike a spherical wave, the intensity of a flat wave does not decrease as rapidly. As a general rule, since sound sources vary so much in size, tones in the bass region have less carrying power, whereas midrange tones, treble tones, and their harmonics retain more of their energy content.

Standing Waves

It is possible for a dry sound to strike a surface and then to be reflected along the line of its arrival. Since their waves will be occupying the same path, there are three possible results. The waves (known as standing waves), direct and reflected, may be in phase, in which case the resultant amplitude will be the sum of the amplitudes of both waves; they may be out of phase, with cancellation as a result; or they may be partially in and partially out of phase. The final result will depend on the location of the listener. The point at which the two waves cancel is a *node*; where they reinforce is known as an *antinode*, or loop. The disadvantage of standing waves is that they can result in dead sound spots in a listening room or unusually loud sounds. The distances of dead spots, or maximum sound spots, are based on the wavelengths of the reflected sound. The nodes are

$\frac{1}{4}$, $\frac{3}{4}$, $\frac{5}{4}$ . . . wavelength from the reflecting surface, whereas the antinodes are $\frac{1}{2}$, 1, 2, . . . wavelengths from the surface.

Standing waves can also be produced by various harmonics that accompany a fundamental wave. Acoustically, standing waves can be minimized or eliminated by having walls that are not parallel, by a multilevel ceiling, or by attaching sound diffusers to the walls and ceiling. A sound diffuser is a panel having an uneven surface. A diffuser having a curved surface is known as a *splay*. A curved surface is desirable, since the angle of incidence of a sound wave will not be equal to the angle of reflection.

Determining the Wavelength of Standing Waves

The distance between the nodes of a standing wave is one-half wavelength. Assume, for example, that a pair of parallel walls are separated by a distance of 10 feet. When a sound wave strikes both of these walls, it establishes nodal points. Thus, these two nodal points are 10 feet (3.0480 meters) apart. Since this is one-half wavelength, a full wavelength equals $3.0480 \times 2 = 6.096$ meters. To obtain the frequency, divide the velocity of the sound (330 meters/s) by 6.096, giving 54 Hz.

Standing waves will be maximum at a frequency of one-half wavelength of the incident sound.

Sound Reflections in Small Rooms

In a small room, especially one having hard, reflective surfaces, multiple reflections are common. As a result there are a variety of interference patterns. Instead of being diffused, various sound frequencies and their harmonics may undergo emphasis, leading to uneven coloration.

Positive Feedback

In a hall or auditorium using a sound-reinforcement system, there is always the possibility that a sound wave from a speaker will reenter a microphone in time to be in phase with existing sound. This can happen when the mic and the loudspeaker are fairly close to each other and when the mic is an omni type. The result is uncontrolled howling. Known as *positive feedback*, it can be cured by moving the monitor (speaker) a greater distance from the mic, by using a cardioid mic with the tail of the mic pointed toward the speaker, and by lowering the gain of the amplifier. In some cases the amount of feedback is not enough to maintain howling, but the feedback is still enough to affect the sound, resulting in sharpness. It is also possible for the sound in a reinforcement system to go in and out of feedback, thus producing sound instability.

A basic sound-reinforcement system consists of a mic, an audio voltage/power amplifier, and a loudspeaker. The sound resulting from this setup should be inconspicuous—that is, the audience should get the impression of sound coming from the musical instruments and not from the loudspeakers.

In some sound-reinforcement systems the monitor speakers are at some distance from the microphones and so the chances for positive feedback are very small. However, in electrical form, the sound from the mics that passes through the speakers is practically instantaneous, whereas the sound from a performer must travel through the air at the much lower speed of approximately 344 meters/sec. The result is that members of an audience, depending on their seating locations, can hear the same sound twice. To overcome this fault, sound-reinforcement systems in large auditoriums or halls may use electronic time-delay circuitry following the audio power amplifiers, so the sound from the speakers and the direct sound are in phase.

THE SABIN

The *sabin* is a unit of sound absorption, representing the absorption obtained from 1 ft^2 of a perfectly absorptive surface. An open window could be regarded as such a surface, with materials such as walls, ceilings, and floors having greater or lesser degrees of sound absorption. There are two main effects when sound impinges on a surface: absorption and reflection. The Sabine formula states that the reverberation period, expressed in seconds, is equal to one-twentieth of the volume of a room in cubic feet divided by the total absorption units for sound in the room.

Reverberation Time

The time required for reverberant sound to decrease by 60 dB, or one-millionth of the original sound-source intensity (assuming the original sound has ceased or has changed), is referred to as *reverberation time*, and it does not take frequency into consideration. The formula for reverberation time was developed by Professor Wallace C. Sabine in 1895 and is expressed as

$$T = \frac{0.05V}{a}$$

where T is the reverberation time in seconds, V is the volume of the room in cubic feet, and a is the total equivalent of sound absorption in sabins per square foot of surface material.

Strength of Reflected Sound

When sound moves through the air, it surrenders some of its energy to the air molecules it encounters with a transfer of energy in the form of heat energy. As a result, the total sound energy generated at the source by instrumentalists and vocalists never reaches the various reflective surfaces. Thus, the sound energy that is reflected never has the strength of the dry sound, assuming there are no resonances in the auditorium or room.

The wet-sound loss is dependent on the area and the materials of which the reflecting surfaces are made, but frequency is also a factor. Treble tones suffer a greater loss than the lower-frequency midrange and bass tones. Although the frequency dividing line is not a definitive one, most of the sound energy is dissipated above 1500 Hz.

Decay Time

The *decay time* is the amount of time sound energy takes to fade out (*decay*) in a room (due to absorption in various materials present in any room) and is directly related to the size of the room. Decay time is expressed in seconds and specifies the theoretical time it takes the sound pressure level to decrease by 60 dB after the input signal has been cut off.

The basic Sabine formula can be rewritten to take into account the conversion of sound energy to heat energy in its passage through the air. Using the metric system, the formula can be written as:

$$T_{60} = \frac{0.16V}{A(A + aV)}$$

In this formula, the reverberation time, T_{60}, is in seconds, V is the volume of the room in cubic meters, A is the total amount of sound absorption in square-meter sabins, and a is the absorption.

Determination of Total Sound Absorption

To calculate the total sound absorption of a room or hall, the area of every reflecting surface must be calculated. This is not too difficult with large, fixed-position areas, but when an audience is involved, the value can be only an estimate. There are two extremes: the complete absence of an audience and a condition in which every seat is occupied. Once a total estimate is determined, a more reasonable figure can be calculated by knowing the percentage of occupancy (Table 7-2).

To determine the amount of sound absorption, each reflective area must be multiplied by its coefficient of absorption. The overall absorption is the sum of each value, to which must be added the absorption supplied by the audience, rugs and drapes. An open window does not supply any absorption, and so the sum of the areas of the open portions must be deducted from the total area.

Sound absorption can also be subjective. A long reverberation time suggests a large auditorium, a church or cathedral, or a spacious concert hall. Conversely, a short reverberation time conveys the idea of chamber music or a small room.

TABLE 7-2   ABSORPTION OF SEATS AND AUDIENCE (AT 512 HERTZ)

	Equivalent Absorption (in sabins)
Audience, seated, units per person, depending on character of seats, etc.	3.0–4.3
Chairs, metal or wood	0.17
Pew cushions	1.45–1.90
Theater and auditorium chairs	
Wood veneer seat and back	0.25
Upholstered in leatherette	1.6
Heavily upholstered in plush or mohair	2.6–3.0
Wood pews	0.4

Acceptable Reverberation Times

Acceptable reverberation times, the amounts of reverberation we tend to associate with rooms or large enclosures, are shown in Figure 7-3. These reverberation times are indicated by the shaded area across the graph. As an example, to determine the reverberation time of a hall having a volume of 60,000 ft³, locate the number 6 on the bottom horizontal line. This point represents 60,000 ft³. Move upward along this line to the bottom of the shaded portion. The line intersects the shaded portion at a value of slightly more than 1.0 s. Continue upward along the same line until you come to the top of the shaded area, for a reverberation

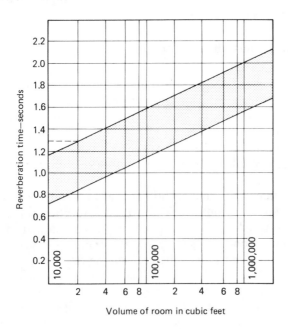

Figure 7-3   Possible reverberation times. The shaded area shows those that are acceptable.

time of about 1.5 s. Thus, the desirable reverberation time of this hall is somewhere between slightly more than 1.0 to slightly more than 1.5 s.

The formula for reverberation time is derived empirically from experimental efforts and does have limitations. In general, it is suitable in situations in which the sound field consisting of dry and wet sound is uniformly distributed. It gives best results when the enclosure has uniform surfaces not complicated by irregular shapes. Other problems include the fact that the coefficient of absorption, as supplied may not be a single number but may be stated as a range. Further, different values of the coefficient of absorption may be given in the literature. Manufacturers of building materials do not have a uniform, standardized method of testing for absorption or may use different equipment for doing so. There is no specific frequency that has been selected for making coefficient-of-absorption measurements, although 500 Hz and 1 kHz are commonly used.

The formula used for the calculation of reverberation time previously supplied does not take the absorption coefficient of air into consideration. This absorption coefficient is significant at frequencies above 1500 Hz. At higher frequencies the formula for the calculation of reverberation time (using the metric system) is

$$T_{60} = \frac{0.16V}{S + a_aV}$$

where T_{60} is the reverberation time in seconds, V is the volume in cubic meters of the room, recording studio, or concert hall being measured, S is the total absorption in square meter sabins and a_a is the absorption coefficient of air. For air, with an assumption of a relative humidity of 50%, the absorption coefficient is 0.30 at 1000 Hz, 0.90 at 2000 Hz, and 2.40 at 4000 Hz.

Arrival Time of Sound

Figure 7-4 is a generalized graph depicting the time of arrival of reverberant sound. The arrival time depends on the positioning of the listener in the audience. Note also that reflected sound has a smaller amplitude than direct sound. Further, direct

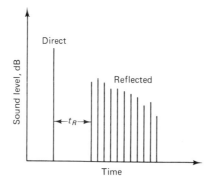

Figure 7-4 Arrival time of reverberant (reflected) sound.

sound is represented by a single vertical line. Reflected sound consists of a number of vertical lines, indicating that the sound behaves as though produced by multiple sources. It is actually an acknowledgment that the sound consists of a number of reflections. These are separated timewise but should be so close to each other that the listener is unaware of the separation.

The time of arrival of dry sound and wet sound at the ears of a listener (Figure 7-5) can be stated as

$$t_D = \frac{R_D}{v}$$

$$t_R = \frac{R_1 + R_2}{v}$$

where t_D is the time of arrival of dry sound; t_R is the time of arrival of wet sound, v is the velocity of sound; R_D is the distance traveled by the direct sound, and R_1 and R_2 are the distances traveled by the wet sound. Dry sound takes a direct path and so has a shorter path than wet sound. Although the drawing shows just a single sound reflection, it is possible for sound to be reflected two or more times. The distance traveled by wet sound is the sum of the distances of the various reflections.

$$t_I = t_R - t_D = \frac{R_1 + R_2 - R_D}{v}$$

If the direct and reflected distances are calculated in meters and if we consider the velocity of sound to be 345 meters/second when measured at room temperature, then the time it takes for the composite sound to reach a listener is expressed by

$$t_I = \frac{R_1 + R_2 - R_D}{345}$$

Environmental Statistics

One way to become acoustically knowledgeable is to visit concert halls, jazz clubs, churches, motion picture theatres, and recording studios and, if possible, to mea-

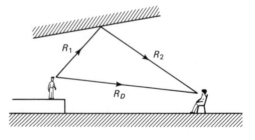

Figure 7-5 Times of arrival of dry and wet sound at the ears of a listener.

sure the acoustic characteristics that differentiate one environment from another. A simulator can be obtained by using a digital sound field processor.

Coefficients of Sound Absorption

Open areas, such as an open window or the spaces in an open-air amphitheater have a sound absorption of zero. Such a value indicates no sound absorption at all. At the other extreme is a material that absorbs all sound incident upon it. The coefficient of sound absorption in this case is 1, or 100%. The coefficients of sound absorption of other materials are somewhere between 0 and 1. These two digits represent the limits. Most substances are partially reflective and partially absorptive.

Table 7-3 lists the sound absorption coefficients for various building materials.

A concrete floor has a sound absorption coefficient of 0.015, or 1.5%. This indicates that the floor will reflect 98.5% of incident sound ($1 - 0.015 = 0.985 = 98.5\%$). This is to be expected of a hard, dense material such as concrete. The data in Table 7-3 do not take frequency into consideration.

TABLE 7-3 SOUND ABSORPTION COEFFICIENTS OF VARIOUS BUILDING MATERIALS

Material	Sound Absorption Coefficient
Linoleum on concrete floor	0.03 to 0.08
Upholstered seats	0.05
Ventilating grilles	0.15 to 0.50
Painted brick wall	0.017
Unpainted brick wall	0.03
Unlined carpet	0.20
Carpet	0.37
Light fabric, 10 oz per square yard	0.11
Medium fabric, 14 oz per square yard	0.13
Medium velour	0.75
Heavy fabric, 18 oz per square yard	0.50
Heavy drapes	0.72
Concrete or terrazo floor	0.015
Wood floor	0.03
Glass	0.027
Marble	0.01
Plaster on brick	0.025
Plaster	0.03
Rough finish plaster	0.06
Wood paneling	0.06

Effect of Frequency on Sound Absorption

Depending on the material, the sound absorption coefficient will vary with frequency. With some substances, such as concrete, the effect is slight; with some substances, there is an increase in absorption, whereas in others, there is a decrease. Table 7-4 shows that $\frac{3}{16}$-in. plywood falls into the latter category.

TABLE 7-4   VARIATION OF SOUND
ABSORPTION COEFFICIENT WITH
FREQUENCY FOR $\frac{3}{16}$-IN. PLYWOOD

Frequency (Hz)	Sound Absorption Coefficient
125	0.35
250	0.25
500	0.20
1,000	0.15
2,000	0.05
4,000	0.05

Table 7-5 indicates the effect of frequency on sound absorption coefficients for a variety of materials.

Determination of Total Sound Absorption

The total sound absorption in a studio or in a home can be determined by listing the areas of all reverberant surfaces and then multiplying those areas by their sound absorption coefficients. A more accurate result can be obtained by taking frequency into consideration.

Example

What is the total sound absorption in sabins of a wooden floor that measures 10 ft × 20 ft? The floor is bare and is not covered by a rug.

Solution   The area of the floor is 10 ft × 20 ft = 200 ft². Assuming a frequency of 1 kHz, its sound absorption coefficient at that frequency is 0.07 based on the data supplied in Table 7-5. Multiplying the area by the sound absorption coefficient gives 200 × 0.07 = 14 sabins.

The same procedure can be followed for all the other surfaces and materials used in the room. The total supplies the sound absorption in sabins of the room. In doing calculations for a room, use the same frequency for each material.

To determine the total absorption in sabins for different frequencies, repeat the arithmetic for frequencies such as 125, 250, 500, 1,000, 2,000, and 4,000 Hz. The result can be plotted in graph form, with frequency indicated along the X-axis and total sound absorption in sabins along the Y-axis.

This work is essential to a prior determination of reverberation time.

TABLE 7-5 SOUND ABSORPTION COEFFICIENTS OF COMMONLY USED MATERIALS AT VARIOUS FREQUENCIES

Material	Frequency in Hz					
	125	250	500	1,000	2,000	4,000
Glass window, closed	0.35	0.25	0.18	0.12	0.07	0.04
Lightweight drapes	0.03	0.04	0.11	0.17	0.24	0.35
Heavy drapes	0.14	0.35	0.55	0.72	0.70	0.65
Wood floor	0.15	0.11	0.10	0.07	0.06	0.07
Carpet (on concrete)	0.02	0.06	0.14	0.37	0.60	0.65
Brick, glazed	0.03	0.03	0.03	0.04	0.05	0.07
Carpet, heavy, on 40-oz hairfelt or foam rubber	0.08	0.24	0.57	0.69	0.71	0.73
Fabric, medium velour, 14 oz per square yard, draped to half of its flat, undraped area	0.07	0.31	0.49	0.75	0.70	0.60
Floors						
cement or terrazzo	0.01	0.01	0.015	0.02	0.02	0.02
linoleum, asphalt, rubber, or cork tile on concrete	0.02	0.03	0.03	0.03	0.03	0.02
wood	.15	.11	.10	.07	.06	.07
Glass						
large panes of heavy plate glass	0.18	0.06	0.04	0.03	0.02	0.02
Gypsum board, $\frac{1}{2}$ in., nailed to 2-in. × 4-in. lumber, centers 16 in. apart	0.29	0.10	0.05	0.04	0.07	0.09
Marble or glazed tile	0.01	0.01	0.01	0.01	0.02	0.02
Plaster, gypsum or lime, rough finish or lath	0.02	0.03	0.04	0.05	0.04	0.03
Plywood paneling, $\frac{3}{8}$-in. thick	0.28	0.22	0.17	0.09	0.10	0.11
Water surface, as in a swimming pool	0.008	0.008	0.013	0.015	0.02	0.025

As an additional example, the dimensions of a hall are 112 × 50 × 28 feet and the volume is approximately 175,000 cubic feet. Using the data supplied in Table 7-5, the absorption coefficients are listed in Table 7-6. A frequency of 1,000 Hz is assumed but other frequencies can be tried as well. The areas are multiplied by the absorption coefficients resulting in the equivalent absorption values in sabins. The total absorbing power of the bare room is 1,146.9 sabins to which is added the absorption of 800 bare seats, from Table 7-2, supplying a final total of 1,266.9 sabins.

Calculation of Reverberation Time

It is possible to calculate the reverberation time of a studio, concert hall, or listening room in the home once the sound absorption has been determined. This can be a single calculation for a single selected frequency, or it can be repeated for a selected number of frequencies.

TABLE 7.6   CALCULATION OF THE SOUND ABSORPTION OF A HALL

Material	Dimensions (ft)	Area (ft^2)	Absorption Coefficient	Equivalent Absorption (sabins)
Floor, cement	56 × 112	6,272	0.02	125.4
Walls, plywood/panel	8 × 336	2,688	0.09	241.92
Plaster	20 × 336	6,720	0.03	201.6
Ceiling, plaster	56 × 112	6,272	0.03	188
Curtain, velour	39 × 20	780	0.5	390
Total absorbing power, bare room				1,146.9
Plus 800 unupholstered seats at 0.15 sabin				120
Total absorbing power, on one present				1,266.9

The first line in Table 7-7 is for a hall having seats that are not occupied. The greater the occupancy, the greater the amount of sound absorption and the lower the reverberation time. Since seat occupancy can be a variable, one method is to keep a record of seat use and then to establish an average occupancy. Depending on the individual, the absorption in sabins for a person will be in the range of 2 to 8.

TABLE 7.7   EFFECT OF AN AUDIENCE
ON REVERBERATION TIME

Audience (number present)	Absorption (sabins)	Reverberation Time (s)
0	1,201	7.3
200	2,011	4.3
400	2,821	3.1
600	3,631	2.4
800	4,441	2.0

Table 7-7 shows the effect of an audience on reverberation time. Note that doubling the occupancy does not multiply the absorption in sabins by a factor of 2. Absorption increase is not linear.

Effect of Excessive Reverberation

Absorption and reverberation time vary somewhat inversely: the smaller the absorption, the greater the reverberation time. Small absorption can be overcome by adding more-absorbent materials to the walls or by adding acoustic tiles to a ceiling that has a plaster reflecting surface.

If, for example, the reverberation time has been calculated as 4 seconds, a desirable alternative might be 2 seconds. To determine the required absorption in sabins, use the formula for reverberation time, with the unknown value being S_a.

Example

Assume a hall has a volume of 150,000 ft^3 and the desired reverberation time is 2.0 s.

Solution   Using the customary system,

$$T_{60} = \frac{0.05V}{S_a}$$

where T_{60} is the reverberation time in seconds, V is the volume of the room or hall in cubic feet, and S_a is the total amount of sound absorption in square-feet sabins. Transposing T_{60} and S_a gives

$$S_a = \frac{0.05V}{T_{60}}$$

Utilizing the data supplied,

$$S_a = \frac{0.05 \times 150,000}{2.0}$$

$$= \frac{7500}{2.0} = 3,750 \text{ sabins}$$

Since different materials will have different coefficients of absorption, it will be necessary to decide which will be used and also the area to be covered by the material. To determine the area of treatment, apply the following formula:

$$\frac{\text{Required absorption in sabins}}{\text{Absorption coefficient}} = \text{area in square feet}$$

It now becomes necessary to determine the material to be used. Select a material that has a high absorption coefficient. If the floor isn't carpeted, a wall-to-wall carpet could be used. Heavy drapes could also be considered. A medium velour fabric could be used. It is also possible to try a combination of these materials.

Since the sound absorption coefficient varies with frequency, a frequency selection must be made. Consider 1 kHz as a possible average. Using Table 7-3 and arranging the suggested materials in tabular form supplies:

Medium velour fabric	0.75
Carpet	0.37
Heavy drapes	0.72

Since the objective is to reduce the reverberation time from 4 seconds to 2 seconds, the next step would be to double the amount of absorption from 3,750 sabins to 7,500 sabins. Note that the velour fabric and heavy drapes have twice the absorption capabilities of the carpet. In terms of percentage, the contributions by the medium velour fabric and the heavy drapes will each be 40% and the carpet 20%. The total number of sabins required will be:

$$7{,}500 - 3{,}750 = 3{,}750$$

$0.40 \times 3{,}750 = 1{,}500$ sabins supplied by the medium velour fabric
$0.40 \times 3{,}750 = 1{,}500$ sabins supplied by the heavy drapes
$0.20 \times 3{,}750 = 750$ sabins supplied by the carpet

Knowing the required absorption in sabins and the absorption coefficients of each of the three materials, it is now possible to calculate the area in square feet from:

$$\frac{\text{required absorption in sabins}}{\text{absorption coefficient}} = \text{area in sq. feet}$$

For the velour fabric and the heavy drapes:

$$\frac{1{,}500}{0.75} = 2{,}000$$

For the carpet:

$$\frac{750}{0.37} = 2{,}037$$

These area figures are rather large and so a possible solution would be to use acoustic tile for the ceiling and/or cover the walls with a sound absorbent substance.

Echoes

Echoes can exist when there is a relatively substantial time difference between the arrival of direct sound and reverberant sound at a listener's ears. The greater this time difference, the more noticeable the echo. For speech, an echo can make the sound unintelligible; for music, besides being annoying, echoes can make the sound appear to be weak and lacking in richness.

Reinforced Sound versus Live Sound

A reinforced sound system uses one or more microphones and add-on electronic equipment such as audio amplifiers and loudspeakers. Electronic reinforcement is used in recording studios and in concert halls and auditoriums where the sound would be otherwise inadequate. The frequency response of the sound is not only determined by the architecture of the listening environment but also by the quality of the electronic equipment. It cannot be better than the quality of the poorest component. The sound response is also determined by the way in which the performing artist handles the microphone.

Live sound does not make use of electronic sound reinforcement. It is commonly used in small rooms, and what is heard is directly dependent on the vocal power of the performer or performers. In this case electronic equipment can be used for recording purposes but is not involved in augmenting the sound to be heard by an audience.

Listening Room Nonlinearity

A listening room is often included as one of the components of a sound system, as it should be. However, any acoustic environment is nonlinear. It does not have a flat frequency response, since reverberant sound suffers a treble frequency loss. Wet sound has a narrower frequency band than dry. Further, the outer part of the ear, the pinna, is shaped so as to favor direct sound, while hindering the reception of sound from the rear. The quality of sound is also affected by reverberation, which can lower its intelligibility. The same effect takes place in music, with instruments seeming to lack sharp definition.

Sound in a reverberant room is affected by the shape of the reflecting surfaces. A concave surface focuses the sound and is particularly unsuitable for recording situations. A convex shape spreads the sound over a large area, much more so than a flat surface. The extent of sound absorption, then, is dependent on a number of factors: the area of the surface, the depth of the material, the type of material, its shape (whether flat, convex, or concave) and the quality of the surface (whether smooth or rough). The angle of sound incidence can also be significant, since it determines if the sound will strike other reverberant materials or will proceed directly to the ears of the listener.

Acoustics of a Room Used Only for Speech

The prime requirement of a room used only for speech is clarity. Acoustically, the room is a failure if part or most of the audience is unable to distinguish the words of the speaker (or speakers). Part of the problem may be due to the way in which the microphone is used. Often, the user is not given any instructions on how to hold a microphone because of a completely unwarranted assumption that no instructions are necessary.

ARTIFICIAL REVERBERATION

One of the problems of natural reverberation is that it is often difficult to predict prior to the construction of a studio or concert hall. Some of the variables affecting reverberation include windows that can be opened or closed, the number of people in an audience, their distribution, and the kind of clothing they wear, the positioning of the microphones, the kind of orchestra, and the kind of music being played. Some of these variables can be controlled, but in a sense an auditorium or recording studio must be "tuned" to achieve best reverberation time. Often, however, the changes are tedious, experimental, and expensive, with no guarantee of success.

Control of natural reverberation can be difficult but can be overcome through the use of artificial reverberation. There are various ways of obtaining artificial reverberation: by a system of vibrating springs, by using a reverberant chamber, by an endless magnetic tape, by physical methods, by analog or digital electronic techniques, by the number and positioning of microphones and the recording artists, by the amount of sound reinforcement, and by the settings of tone, equalizer, and level controls in recording equipment.

The reason for using artificial reverberation is not to create the impression that studio recording has been used but rather to create a feeling of size and reverberation of a variety of architectural arrangements, from small rooms to concert halls. This recognizes that reverberation is part of every recording.

Artificial reverberation is now used extensively in recording studios or broadcasting stations. A reverberation unit can supply two sound channels, which can be used jointly or separately. The decay time of each channel can be adjusted independently and continuously, without creating any noise, with the unit responding instantly to any readjustment of decay time. In a dry studio environment, individually adjusted reverberation can be added to the various tracks of a multitrack recording.

Techniques for Adding Artificial Reverberation

There are quite a few ways to use a pair of reverberation channels.

1. By adding individual reverberation characteristics to each track of a multitrack recording. Decay times and intensities can be adjusted separately for each track, even if reverberation is added to two tracks simultaneously, assuming the use of a two-channel artificial reverberation component.

2. By feeding independently from the two stereo channels and mixing reverberation with the same decay time and intensity onto the original stereo panorama via two pan-potentiometers in order to broaden the original stereo base.

3. By feeding only one signal into the reverberation unit in monaural operation and splitting the reverberation signal into two echo-return signals to be fed into the mixer or the amplifiers. The decay time and reverberation intensity of each channel may be varied independently.

Artificial reverberation is essential in recording studios, broadcast stations and theaters. It can be used experimentally to determine the most desirable amount of reverberation, permitting permanent acoustic changes to be made more scientifically and less haphazardly. However, room reverberation may need to be made variable, under supervised control. The orchestra conductor and attending audio engineers may need to use variable control for particular compositions.

8

Synthesizers

A synthesizer is an electronic device made up of various circuits, including oscillators (electronic generators), filters, waveform (envelope) shapers, and amplifiers. All can be and are controlled by voltages (not by sound). The control element can be a keyboard similar to that on a piano or a foot pedal (similar to that used by a piano). Some synthesizers have a microphone input; many do not.

A basic difference between a musical instrument and a synthesizer is that the former is mechanical; the synthesizer is electronic. With an instrument, the time duration of a tone is limited; with a synthesizer the tone can be continued indefinitely.

A synthesizer resembles a piano somewhat because it uses a keyboard. A piano produces a single note when one of its keys is depressed. It is a percussive instrument and depends on the striking of wire strings. Pedal action can sustain the note produced, soften it, or curtail it. Other than these effects, the note cannot be modified by the piano, and it will always be recognizable as a piano tone.

The synthesizer performs in a similar manner. But unlike other musical instruments, it is entirely electronic. And unlike the piano, it is equipped with switch-type controls positioned near the keyboard that can direct electronic circuitry to shape the selected tone. The synthesizer should be looked on as an entirely new musical instrument capable of producing an astonishing variety of sounds, some of which may never have been heard before—sounds that are a simulation of those generated by a variety of instruments. The variety of new music and modified forms of existing music, often aided by or completely produced by synthesizers, is extraordinary. It is also possible that without the help of synthesizers some of these music forms could not have come into existence. A partial listing includes metal, rap, dance, pop, Latin, reggae, funk, African Zulu, baroque house, rave, avant-garde, jazz, techno-pop, folk-rock, punk, Mexican, Western, Spanish, rockabilly, calypso, including various modifications of these. Some are combinations such as rock-funk hip hop or reggae-dancehall-funk-hip hop. Musical instruments create sounds, and these sounds can be controlled to a limited extent by a musician. The synthesizer also generates sounds, but the amount of possible control is extraordinary.

The performer on a musical instrument such as a piano or a violin can control two important elements of sound: pitch and volume. The timbre is an inherent characteristic of the instrument and is not under the control of the artist. It is the timbre of a tone that enables us to differentiate musical instruments. It is the timbre that lets us recognize a violin, a flute, or a piano. The big difference between an instrument and a synthesizer is that with the synthesizer, the performer—for the first time—can control timbre. The synthesizer is not an artificial piano, flute, or any one of dozens of other instruments. But it can control timbre, supplying either completely new tones or tones that resemble those of other instruments.

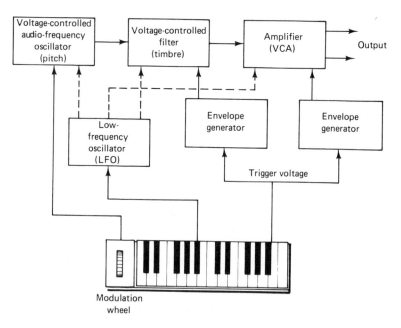

Figure 8-1 Technique of waveform control.

OVERALL VIEW OF THE SYNTHESIZER

Figure 8-1 is a block diagram of a synthesizer. Although this diagram is representative, there are variations among synthesizers. The keyboard is the control element, but it is usually accompanied by various selector controls as well, mounted to one side and above the keyboard. The arrows show the direction of movement of electrical voltages through the system. The output is also an electrical voltage and is supplied to one or more loudspeakers.

SYNTHESIZER OSCILLATORS

The synthesizer has a pair of oscillators, one a low-frequency oscillator (LFO) and the other, a voltage-controlled oscillator (VCO) for determining pitch.

An oscillator is an AC generator producing an output waveform in two frequency ranges: the subsonic, consisting of all frequencies below 20 Hz (and as low as 0.03 Hz), so-called since they are below our hearing range, and audio frequencies produced by oscillators whose frequencies extend from 20 Hz to 20 kHz. The hearing of many people is well within the audio range.

The output of an oscillator is an AC voltage and, depending on its design, can have a variety of waveforms, including the sine, square, sawtooth, triangle, and pulse waveforms. Pulse and square waves are sometimes considered synonymous, but they are different. A square wave is actually that, a wave whose time duration is equal to its amplitude, using its X axis as a reference. The pulse wave is ordinarily DC. Its time duration, as shown by its horizontal segment, is very short compared with its amplitude.

Dual Oscillators

One of the reasons a piano produces beautiful sound is that it is a multistringed instrument. Although these strings are tuned to the same pitch, the sounds they produce are somewhat out of phase.

A similar condition can be produced in a synthesizer by using more than one VCO. The second VCO has a number of advantages. It can make the selected voice sound much more interesting, supplying a richer, fuller tone.

The Low-Frequency Oscillator (LFO)

The LFO supplies an output signal that has a very low frequency. It can be used to supply a driving signal voltage to the second of the two oscillators, the VCO. The signal can also be used as a driving voltage for a following voltage-controlled amplifier.

The waveshape of the subaudio output of the LFO can be a sine wave, a square wave, or a triangle wave and is referred to as *modulating voltage*. The modulation is used to give the final tone a sound that is continuously repetitive.

The effect of the use of the LFO depends on two factors: the kind of waveform it has and the ultimate destination of that waveform. Thus, sine-wave modulation of the VCO results in vibrato; using square-wave modulation input to the VCO supplies a trilling effect; tremolo is obtained by sine-wave modulation of the voltage-controlled amplifier (VCA); and square-wave modulation of the same circuitry produces an effect that resembles an echo.

Still another effect can be obtained with the LFO when it uses a square-wave modulating waveform, which is an automatic variation of the pulse width of the VCO. This results in a sound that is richer and fuller.

Oscillator Voltages

An oscillator or sound generator in a synthesizer requires two types of DC voltage. One of these is an operating voltage. This voltage is to the oscillator what gasoline is to a car—a source of energy. The driver of the car uses an accelerator pedal to control the speed of the car. The oscillator uses a DC voltage for controlling the oscillator's frequency. The output of the oscillator is a sine wave, but this sine wave can be modified to have any desired shape, such as a sawtooth wave or a square wave.

Oscillator Frequency

The frequency of an oscillator, the number of cycles per second it produces, is its pitch. In a piano, an increase in pitch is obtained by moving up along the keyboard, from left to right. With a synthesizer a change of pitch is obtained by supplying the voltage-controlled oscillator with a pitch-control DC voltage. The higher the voltage applied to a control element of the oscillator, the higher its output pitch. The fact that the oscillator is voltage controlled is the source of its name, voltage-controlled oscillator.

The Voltage-Controlled Oscillator

The VCO is the second of the synthesizer's two oscillators. This is the oscillator that determines the pitch (that is, the frequency) of the sound output. Pitch is controlled by the amount of input voltage; the higher this is, the higher the frequency. Although the LFO can be used to produce certain effects, it is the VCO that can be considered as the beginning of the waveshape and the resulting voice (waveshape).

OSCILLATOR WAVESHAPE

Depending on the design of an oscillator, its output can have a variety of shapes, and it is these shapes that determine the sound of that output. Despite the available variety, most of the waveforms selected for synthesizer use are fairly standard, consisting of the sine, triangle, sawtooth, trapezoidal, and square waves.

Sine Wave.   The *sine wave* is the simplest of all waves and is characterized by the fact that it has no harmonics. Consequently it is a *pure wave*, but it is not subject to manipulation by a filter. However, it can be changed by sending it through an amplifier, with the shape of the output wave dependent on the amplifier's design and working voltages.

The drawing in Figure 8-2(a) is that of various waveforms starting with a sine wave. The single vertical line above the digit 1 in Figure 8-2(b) represents the amplitude, or maximum positive peak, of such a wave. No vertical lines appear above the numbers that follow, since there are no harmonics. This waveform is used for musical vibrato and tremolo.

Sawtooth wave.   A sawtooth waveform and a harmonic spectrum appear in part (a) and (b). Its sound is similar to that produced by reed instruments (such as saxophone). The synthesizer also uses it to generate the tones of sustained string instruments. It is characterized by having both odd and even harmonics, with these having decreased amplitudes. The total number of harmonics is important, since the greater their number, the richer and fuller the sound. This wave and any others that do not have the shape of a sine wave are known as *nonsinusoids*.

Phase angles do not apply to nonsinusoidal waveforms, since angular measure is applicable only to sine waves. The X axis of the sine waveform is marked in degrees, a condition not found in connection with nonsinusoids. The slant portion of the sawtooth increases linearly and is referred to as a ramp voltage. The sawtooth is sometimes called a ramp wave.

Rectangular wave.   Figure 8-2(c) is that of a *rectangular waveform*, also known as a *pulse wave*. A single complete cycle extends from point a to point b. Note that the positive and negative half-cycles are symmetrical in amplitude but not in time. The harmonics of this waveform are both odd and even order, but it can have missing harmonics.

The positive and negative alternations of the pulse wave (d) need not necessarily have the same amplitudes. This means that the waveform can have two voltage states, with one larger than the other. The amount of time used by the pulse in the higher of its two voltage stages is referred to as its *duty cycle*.

Drawing 8-2(d) in Figure 8-2(a) shows another type of rectangular wave that is symmetrical in both amplitude and time. This wave has odd-order harmonics only as in Figure 8-2(b) part (d).

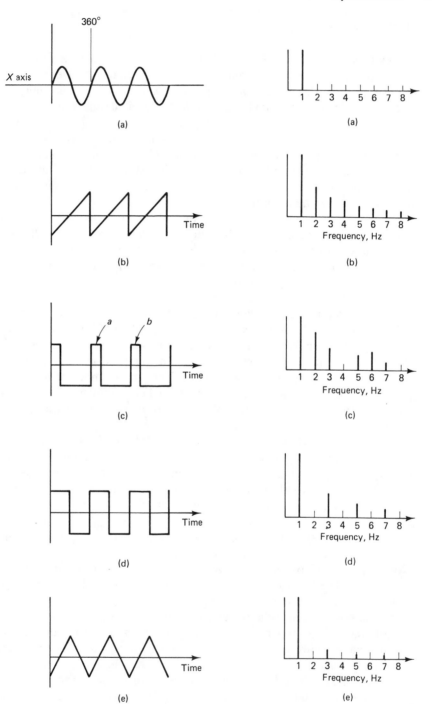

Figure 8-2a Possible oscillator output waveforms.

Figure 8-2b Instantaneous amplitudes of waveforms of the type in Figure 8-2a.

Triangular wave. Figure 8-2(e) in Figure 8-2(a) is a triangular wave. It does have some resemblance to the sawtooth, but it has only odd-order harmonics, and these have very small amplitudes. Consequently, most of the energy is contained in the fundamental, with very little in the harmonics. Its sound is similar to that of a trumpet.

A synthesizer can contain one or more oscillators, which are actually signal generators. Oscillators can be considered as the heart of a synthesizer and are controlled by DC voltages. In turn, it is the DC voltage-controlled oscillators (VCOs) that trigger the following circuitry. The extent and duration of the DC voltage can be determined by a keyboard, by a sequencer, or by various devices mounted above the keyboard. These consist of switches or some form of rotary control, such as a variable resistor or some combination of the two.

Voltage control is applicable not only to the VCO but also to the following voltage-controlled filter (VCF) and voltage-controlled amplifier (VCA). The separate control voltage circuits are represented by the letter X in Figure 8-3.

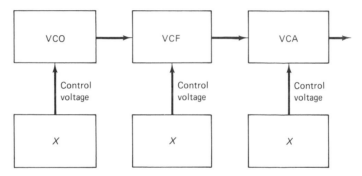

Figure 8-3 The VCO, VCF, and VCA can be independently controlled.

Analog versus Digital Oscillators

All synthesizers, analog or digital, start the generation of sound with an electronic oscillator. The output of the oscillator is analog, and from that point on it can remain analog throughout the synthesizer, in which case it is referred to as analog. However, there is an option of converting that analog form to digital, and in that case it is known as such. Ultimately, though, to be heard by human ears, all digital forms must be converted to analog. So, every digital synthesizer is actually a combination of digital and analog.

Computer Application

Digital oscillators and computers work in the same way in the sense that both use binary arithmetic, consisting of only two digits, 0 and 1. A number of these digits in succession is called a bit stream, and that bit stream can be used to represent music or, more precisely, pitches of sound. The sounds can be generated by the synthesizer oscillators, controlled by either a computer keyboard or a specially adapted piano keyboard. Ultimately, though, the digital bit stream must be fed into a digital-to-analog circuit (D/A) for conversion into analog form. It can then be heard through headphones or speakers.

The digital oscillator has a number of advantages over the analog. A musician can use it to create completely new waveforms, and these can be manipulated and recorded. The sounds that are produced can be completely new, never heard before. Still another advantage is that because of operational similarities between the digital synthesizer and the computer, both can become associated with the computer controlling the synthesizer.

The frequency of the analog oscillator can be changed by altering the values of its electronic components, but an easier and more controllable method is through the use of a controlling voltage. Thus, the frequency of an oscillator might require a change of 1 V to produce a change of 1 octave. This makes two assumptions: The first is that the oscillator will change frequency by 1 octave, and the second is that each succeeding voltage will be precisely 1 V.

The waveform output of the analog oscillator remains fixed, but it can be changed by subtractive synthesis, a waveform-shaping circuit. The assumption is made that the shaping circuit performs equally and uniformly well throughout the entire audio range. To the extent that they do not do so, usually at the lowest and highest frequencies, the voice envelope can be different. In effect, it is the timbre that changes, and it is as though a violin decided to change its shape, resulting in a sound that is somewhat different, depending on the frequency of its output.

The waveform of a digital oscillator is generated by additive synthesis. The waveforms are produced by the successive closing of electronic switches working in a predetermined sequence. No smoothing filters or shaping networks are required.

Oscillator Noise Generator

The synthesizer may include an oscillator noise generator (Figure 8-4), a sound source for producing wind, thunder, percussive effects, and other sounds, depending on the make and model of the unit. It can also produce white noise, containing all the frequencies in the audible range but not having a periodic pattern. The noise generator (Figure 8-5) can produce random triggers and pitches when used in combination with other synthesizer functions. Generally, the only control for a noise generator is one used for adjusting its volume. Some of the more sophisticated synthesizers give the user a choice of pink noise or azure noise in addition to white noise.

Since the noise generator is a nonperiodic device—that is, it supplies sounds of all frequencies—the filter is a wide band-pass type. Following the filter is the circuitry (the VCA) used to amplify the noise signal, and it is to the VCA that the control-signal voltage is applied. The resulting sound will depend on the way

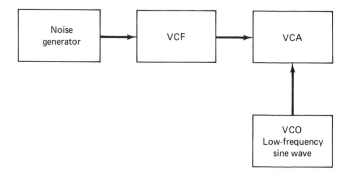

Figure 8-4 A noise generator is not a control voltage but is an additional sound source in the form of a voltage. The VCO patched into the VCA adds vibrato. Note that an alternative patch would be to have the VCO connected to the VCF.

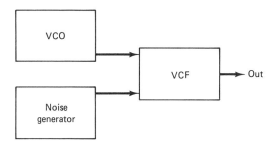

Figure 8-5 Noise generator setup.

the VCO is used. The voltage it supplies can be continuous or divided into regularly recurring segments, and in this way a variety of noise effects can be achieved. It all depends on the output waveform of the VCO. Different noise effects are obtained, depending on whether the waveform is a sine wave, pulse, square wave, etc.

The Master Oscillator

The *master oscillator* is the first of the two VCOs (Figure 8-6). It can produce a pitch proportional to the setting of its range switch and tuning control. It covers not only the audio range but also subsonic frequencies, those frequencies starting at 20 Hz and going down as low as a fraction of a cycle. The output waveshapes can be sine, triangle, or square.

Some synthesizers have as many as four VCOs, which offer six waveforms: triangle, square, sawtooth, inverted sawtooth, pulse, and sloped square. Some of these oscillators may have continuously adjustable waveforms. One or more of the oscillators may have an upper frequency of as much as 40 kHz.

The oscillator frequency is controlled by the keyboard, which determines the type and amount of variation in pitch. The variations may include envelope

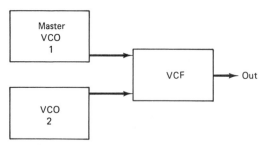

Figure 8-6   Pair of VCOs patched into a VCF.

selection, which produces slow or rapid sweeps up and down in pitch with each key depression. The type of sweep is dependent on the settings of the envelope's attack, decay, and sustain controls.

Oscillator selection produces repeating changes in pitch. The type of change is determined by the oscillator waveform selected. The master oscillator can be used to produce vibratos of different speeds and depths. It can use a sampler to generate a series of ordered or random pitches and an envelope generator to produce sweeping pitches. It can use more than one keyboard to produce discrete pitches having a fixed interval. With some synchronizing units, the pitch of the VCOs can also be adjusted manually.

THE KEYBOARD

The *keyboard* (Figure 8-7) resembles that used on the piano, although it often has fewer octaves and in some instances the keys are smaller in size. The keyboard is used to select the amount of DC voltage for triggering the VCOs in the synthesizer and, in some instances, voltage-controlled amplifiers (VCAs) and filters as well. In addition to its triggering action, a keyboard can also be used for pitch

Figure 8-7   Digital synthesizer. (Courtesy Yamaha Corporation)

control. This is quite unlike a piano, in which the keys are basically used for turning sound on and off. The synthesizer keyboard is also used for shaping sounds.

Bending

The keyboard may be accompanied by a rotary dial positioned either above or to one side, which is used for slightly increasing or decreasing the pitch of a tone, called *bending*. Whether this is needed or not depends on the stability of the controlled oscillators. If they drift in frequency, however slightly, the result is *note detuning*. Oscillator stability is a function of the electronic design and the quality of the parts used in its manufacture.

Keyboard Sensitivity

The keys on some keyboards, usually of the less expensive types, are simply on/off switches and can be considered go–no go devices. Other keyboards, on higher-quality synthesizers, are pressure sensitive and so more nearly approach the keying action of a piano.

Monophonic versus Polyphonic Keyboards

Some keyboards are *monophonic*; with these only a single key can be played at one time. Other keyboards are *polyphonic*; with these more than one key can be played at the same time.

The Split Keyboard and the Layered Keyboard

A split keyboard (Figure 8-8) is one that is *polytimbral* (or *multitimbral*). The suffix *timbral* is taken from the word timbre, the quality given to a sound by its overtones, or the distinctive tone of a particular instrument. In a split keyboard

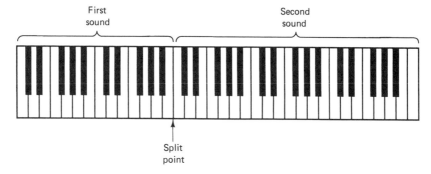

Figure 8-8   Split keyboard. Some keyboards have more than one split point.

one section is used for one timbre, and the other section is used for a different timbre. With a polytimbral keyboard, a performer can play one or the other of the keyboard sections or both simultaneously. Not all such keyboards are alike. Some are programmable—that is, the performer can determine the location of the split point. And some keyboards can be split into more than two sections. Obviously, full-length, 88-note keyboards are more easily adapted for multiple splits.

An alternative keyboard type is shown in Figure 8-9. The keyboard used here is a *layered* type. This is one in which more than one voice (one instrument) can be played at the same time.

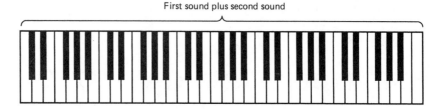

Figure 8-9   Layered keyboard.

Miscellaneous Control Devices

In addition to piano keys and rotary controls, there are various other devices that are used, including foot pedals, push buttons, joysticks, and panels that are touch sensitive.

Keyboard Control

A keyboard resembling that used on a piano is one of the most common techniques for controlling the VCO of a synthesizer. With this, it is possible to play a chro-

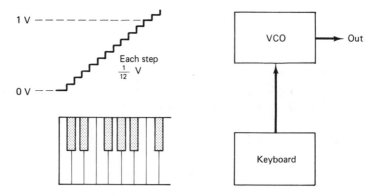

Figure 8-10   Each subsequent key means an additional $\frac{1}{12}$ - V input to the VCO.

matic scale, a scale consisting entirely of half steps. As the key that represents the beginning of an octave is played, the result is a tone, and with each succeeding key, the pitch of that tone is increased by a specific amount. As indicated in Figure 8-10, each key represents the addition of $\frac{1}{12}$ V to the VCO. Since 12 tones are used to form a complete octave, playing 12 successive keys will supply an increasing amount of voltage. In other terms, it requires 1 V divided into 12 parts to produce one octave.

THE VOLTAGE-CONTROLLED AMPLIFIER

The VCA is an audio amplifier whose output signal voltage is governed by a control voltage instead of a variable resistor, such as a volume control. The input control voltage—that is, its signal input—is supplied by a VCO but it could also be any signal that is to be processed, including noise voltages or an envelope generator.

In synthesizers, the control voltage is usually an envelope generator, that is, the instantaneous summation of a large number of audio voltages. Sometimes the control voltage is a low-frequency oscillator, which is used to produce a tremolo effect.

WAVE SHAPING

The tone to be produced by a synthesizer begins with a VCO whose voltage output can be a waveform such as a sawtooth. With the use of a formant filter, the sawtooth can be changed into a specific tonal waveform. This waveform is a complex type consisting of a fundamental and a number of harmonics.

It is generally assumed that harmonics have a smaller amplitude than the fundamental, and although this is correct for a number of tones, it is not valid for all. String and reed voices, for example, have harmonics whose amplitudes are greater than their fundamental waveforms. In the case of the flute, though, it is the fundamental that is dominant, as indicated in Figure 8-11.

The natural strength of harmonic amplitude can be modified in several ways. One of these is by sending the waveform of the voice through an electrical filter. Another is by having the voice adjacent to some physical object that is resonant at a specific harmonic frequency. In either instance, the tonal waveform that is heard is dependent on the effect these actions have on the selected harmonic.

Harmonic modification can be obtained not only through the use of filters but also by circuits known as traps and by ringing circuits. A ringing circuit consists of reactances such as inductors and coils that behave like tuned circuits at the harmonic frequency. A specific harmonic or harmonics, can be not only emphasized but also can be attenuated. These modifications change the resultant waveform and give a particular tone sonic characteristics it may not normally have.

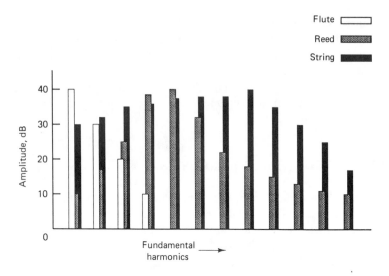

Figure 8-11 Harmonic content of flute, reed, and string instruments.

Ringing

The effect of a ringing circuit depends on how it is made. If it has a high ratio of reactance to resistance, it is capable of generating an ongoing, repetitive wave having high amplitude. If the ringing is strong enough, the effect is a tone that simulates that of a drum and is known as a *percussion waveform*. This periodic wave, if repetitive, is further described as a repeat percussion type.

To obtain a desired voice, it is necessary to generate not only the fundamental sine wave but also a number of harmonics that have a correct amplitude relationship to that fundamental. An envelope (Figure 8-12b) represents all the

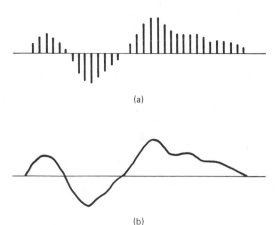

Figure 8-12 When peak points of the instantaneous values (a) are connected, the result is the waveform of a sound (b).

instantaneous voltage values of a wave. Graphing all these values will supply the waveform envelope when all of the instantaneous values are connected across their tops.

A waveform generator controls the attack and release characteristics of a variable voltage. Older synthesizers made use of an AR type, where AR is an acronym for attack release.

A much more desirable form of envelope generator is one capable of handling four characteristics—attack, decay, sustain, and release—from which the acronym ADSR is obtained. The ADSR envelope generator (Figure 8-13) is more desirable, since each of four parameters can be handled independently, thus supplying more control over the ultimate waveshape.

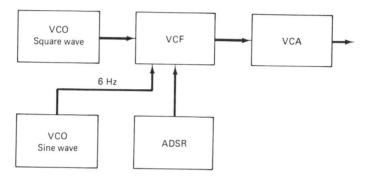

Figure 8-13   ADSR unit is a wave shaper. Sine-wave VCO produces vibrato.

Every musical instrument has its own particular waveform, with the waveform used for instrument identification. The closer the synthesis of the waveform to that produced by a musical instrument, the more nearly the output of the synthesizer generator produces corresponding sounds. The waveforms of a flute and a guitar are completely different. The extent to which the synthesizer can duplicate the waveforms of these instruments is its ability to produce music that sounds like a flute or a guitar.

WAVEFORM SYNTHESIS

Additive Synthesis

Synthesis is the formation of sounds by combining two or more sine waves into complex waveforms. The starting point is a sine wave having a specific frequency; sine waves whose frequencies are multiples of the first sine wave are then added to it. The combining of the fundamental sine wave and separately generated harmonics, accomplished in a device known as a mixer, is referred to as *additive synthesis*, since the starting sine wave and the included harmonics (which are also sine waves) are added to each other.

Harmonics that are added to the fundamental sine wave for the purpose of producing a new waveform must be generated separately. Each of these harmonics has an arithmetical relationship to the fundamental—that is, the frequency of each harmonic is twice, three, or more times that of the fundamental.

The waveform resulting from additive synthesis is complex, since the fundamental and the harmonics are out of phase in various amounts, starting at 0° and extending to 360°. Thus, the various waveforms, fundamental and harmonics, are additive or subtractive during the synthesis. The ultimate shape of the resultant wave is dependent on the number of harmonics in the synthesis.

Additive synthesis can consist of a fundamental plus odd-order harmonics only, a fundamental plus even-order harmonics only, or a fundamental plus both odd- and even-order harmonics. The ultimate shape of the resultant complex waveform depends on which additive arrangement of harmonics is used.

The waveform of any musical instrument can be imitated, resulting in synthesizer sound that corresponds very closely to that of the selected instruments. However, this adds to the complexity of the synthesizer, since it requires a large number of oscillators. If a synthesizer using an additive synthesizer would claim the ability to supply 60 voices, then it would require 60 oscillators, an impractical electronic situation. With additive synthesis, each oscillator would produce a sine wave, which would then need to be suitably modified.

Fourier Transforms

Additive synthesis is based on a mathematical analysis of audio waveforms. The construction of complex waveforms from a basic sine wave plus a number of other sine waves having a selected frequency ratio to the basic sine wave is known as *Fourier synthesis*. It is also possible to start working with a complex waveform and to determine the frequency of the fundamental and those of the added harmonics.

Subtractive Synthesis

The large number of oscillators required for additive synthesis can be reduced through the use of subtractive synthesis. This technique takes advantage of the fact that a complex waveform consists of a fundamental, which is always a sine wave, plus a large number of harmonics, possibly ranging from 2 to 10, all of which are also sine waves. A waveform such as this would be the equivalent of 10 oscillators. Although these harmonics often have amplitudes that are smaller than the fundamental, there are complex sounds in which one or more selected harmonics are larger than the fundamental. Thus, a harmonic of a musical tone could have a very large third harmonic, as an example, when its frequency is that of the resonant frequency of the body of the instrument.

A complex waveform can be modified through the elimination of one or more of the added harmonics, and, in some instances, removing the fundamental frequency as well. Known as subtractive synthesis, this process modifies a complex waveform by sending it through a filter. A number of different kinds of filters are available, including band-pass, band-reject, and low- and high-pass types. When equipped with an IN port, the filters can be controlled with DC voltages, with the triggering action supplied by a keyboard. Less sophisticated synthesizers may have manual controls to determine filter functioning.

Another technique for the construction of sound waveforms is to start with a complex wave and then, through the use of filters, to eliminate parts of the sound. A tone control is one example, but an equalizer is more precise. The filters may be manually or electronically controlled.

DIGITAL SYNTHESIS

The internal functioning of a synthesizer may be analog, digital, or some combination of the two. Clocks are typical examples of analog and digital applications. An analog clock shows all the numbers representing hours, starting with 12 at the top center and followed, in clockwise order, by the numbers 1, 2, 3, and so on. Such a clock may also have a sweep second hand. A digital clock shows only the current hour and minute. With the passage of time, the minute section shows numbers from 0 through 59. Successive hours are displayed one at a time.

A digital synthesizer makes extensive use of a microprocessor. Unlike the analog type, which starts with a waveform, the digital type starts with a series of numbers, with these numbers representing the instantaneous amplitude during

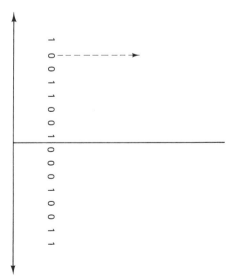

Figure 8-14 Any instantaneous value can be represented in binary form. 16 bits are used here.

each moment of time. Actually, the waveform is the same in both instances. What is different is the way in which it is represented. The digital representation of a wave can be a series of numbers, as shown in Figure 8-14. Note that only two kinds of digits are used: 0 and 1. Other decimal digits are not used in digital representations.

Each digital number can be displayed as a vertical line, with its height dependent on the numerical value of the digital number. When these are assembled, the result is a graph of the wave. When this digital waveform is passed through a filter, the result is a sine wave. A digital oscillator, also known as a digitally controlled oscillator (DCO), can supply the digital equivalent of any other kind of wave—triangular, square, and so on.

SYNTHESIZER CONTROLS

The controls of a synthesizer can be two types. One is the keyboard, arranged in piano keyboard fashion. Like the piano, the keys of the synthesizer control the pitch. The other type is the voice module.

The Voice Module

Controls known as *voice modules* and mounted above the keyboard are used to vary oscillator components in order to supply the equivalent sound of various instruments. Once an instrument type—that is, a specific voice module—has been selected, it can remain fixed until the musician decides to change it. The greater the number of voice modules, the more electronically sophisticated the synthesizer and the more costly it is.

In the meantime, once a particular voice module has been selected, the operator changes the pitch by playing the keyboard just as though it were part of a piano. What we have here, then, are the basic elements of a synthesizer, but the synthesizer is capable of more operating features. Some synthesizers have few features; others have many more. The greater the number of features, the more expensive the synthesizer.

Envelope Control

A synthesizer may have manual controls for the adjustment of the envelope supplied by a particular voice module. These controls can make various changes in the envelope and, in so doing, make modifications in the sound produced by the selected voice.

The Sequencer

A *sequencer* is a device either integrated with a synthesizer or available as a separate component that supplies a series of tones on a one-time basis or repetitively, with those tones remaining in the same order. Thus, the sequencer can operate in two modes: the run mode for continuous operation and the step mode, in which the performer controls each individual sequence.

A sequencer, also called a sequence recorder, is a digital device that supplies commands repetitively, with those commands remaining in the same sequence. The command consists of a set of DC voltages that are applied to the input of a VCO. The sequencer usually functions in connection with a built-in electronic clock working as a timing device for determining the repetition rate of the sequencer's pulses. The data of the digital sequencer are stored in a digital memory. In earlier synthesizers, the sequencer was an analog type, but analog synthesizers required a large group of separate controls, with each of these determining the pitch of a selected tone.

The advantage of a digital sequencer is that its memory not only recalls the order in which the notes are played but their rhythm as well. Sequencers are generally monophonic but can be polyphonic in more sophisticated equipment. With a polyphonic setup, two or more tone sequences can be stored and played individually, sequentially, or in combination.

The sequencer is used to compel the oscillator to produce a steady rhythm. The sequencer may be programmed, so one synthesizer can be used to supply the rhythm automatically and another synthesizer can be keyed for the solo.

SYNCHRONIZATION

If two oscillators drive a voice module, they can be synchronized. Synchronization means that the two oscillators operate on the same frequency. Alternatively, one of the oscillators may be locked onto a selected harmonic of the other, thus affecting the overall timbre of the output. This enriches the sound but results in distortion.

PORTAMENTO AND GLISSANDO CONTROL

The output voltage control of a keyboard moves from one voltage to the next, either increasing or decreasing, in discrete steps, with each transition clean and distinct, an effect that is known as *portamento*. Portamento can be defined as the movement from one tone to another, possibly created by a piano or a bowed string instrument.

Glissando, sometimes incorrectly used synonymously with portamento, is simply a variation; instead of supplying discrete tones, it produces a gradual change between one sound and another.

AMPLITUDE MODULATION

Amplitude modulation (AM) is well known as one of the techniques used for broadcasting a radio signal. A similar technique, not related to broadcast AM, is also a working tool for synthesizers.

In *synthesizer amplitude modulation*, a control voltage is used for producing variations in the loudness of a signal; this is done by driving a voltage-controlled amplifier (VCA) with a voltage signal. The loudness adjustment is made to the beginning and ending of the envelope waveform, thus changing its attack and decay characteristics.

Unlike DC voltages used for triggering VCOs, various voltage shapes, such as sine waves, triangle waves, or pulse waves (square waves), are used. The instantaneous loudness of the output of the VCA follows the general shape of the control voltage. Thus, with sine-wave control, the loudness rises gently to a peak and then decreases just as gently to zero.

The frequency of the control signal helps determine the richness of the sound. The sound that is being modulated is referred to as the *carrier* and the control signal is the *modulating voltage*. With amplitude modulation, upper and lower sidebands are formed. If the control signal is a subaudio type, having a frequency of less than 10 Hz, the resulting tremolo is noticeable.

With a triangle-waveform input, the loudness rises linearly to a peak and then decreases in the same manner. Upon reaching zero, the loudness undergoes a sharp change, increasing once again. With a pulse wave, such as a square-wave type, the loudness rises practically instantaneously to a peak, remains at a fixed level for a predetermined time, and then drops to zero practically instantaneously.

FREQUENCY MODULATION

Another type of modulation used in synthesizers, *frequency modulation* (FM), also has no relationship to broadcast FM. Frequency modulation is used to change the pitch (the frequency) of the sound, achieved by supplying a driving signal to the VCO input. The drive signal has an extremely low frequency, usually in the order of a few hertz. The result is a vibrato, expressed as slight and rapid variations in pitch.

RING MODULATION

A *ring modulator*, supplying a type of amplitude modulation, is a circuit that is driven by two separate audio signals. Working as a mixer, the output of the unit consists of two waveforms, one of which is the sum of the two input signals and the other, their difference. However, the frequencies of the two original input signals do not appear in the output of the modulator.

THE VOLTAGE-CONTROLLED FILTER

The output of a VCO could be supplied, unchanged, to an output port, and from there could go to a tape deck for recording or to an audio amplifier prior to being delivered to headphones or a speaker.

Instead, the voice envelope can be supplied to a voice-controlled filter (Figure 8-15). This component, using various types of filters, works somewhat like a tone control and can emphasize bass or treble tones. If bass tones are to be emphasized, the effect is achieved by cutting down on treble tones. Note that the bass tones are not emphasized or amplified. They sound louder simply because the ratio of bass to treble has been increased.

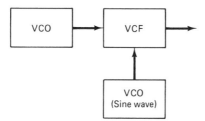

Figure 8-15 Sine wave patched into VCF is used to produce vibrato.

Ordinarily a passive filter produces a fixed result. It may cut off low frequencies, cut off high frequencies, or pass or reject a selected group of frequencies. This action does not change. The *voltage-controlled filter* (VCF) can select any of a group of frequencies and amplify them. It can be said to act like a number of different filters, with all the filters under the control of the performer. By amplifying certain selected frequencies and rejecting others, it alters the timbre of the sound. The frequency changes the quality given to a sound by its harmonic. The voltage control used on the VCF is generally exponential and is expressed in terms of octave per volt; that is, the keyboard generates a $\frac{1}{12}$-V step for each semitone.

The voltage control is established by the keyboard-controlled voltage of the synthesizer. In effect, the filter can be considered as an electronic tone control.

It removes certain audio frequencies from the signal supplied by a VCO. The VCF is often controlled by either an ADSR or AR envelope generator; AR (attack, release generator).

The Band-Reject Filter

Also known as a *notch filter*, the *band-reject filter* works in a manner opposite to a band-pass filter. This filter blocks or rejects a selected spot or narrow band of frequencies referred to as a notch, while passing frequencies above and below it. The notch is sometimes referred to as a *stop band*.

The Cutoff Frequency

The *cutoff frequency* is the point at which the filter is 3 dB down from its maximum. In either the low-pass or high-pass filters, it is the frequency at which filter rejection begins. In the case of a band-pass, or notch, filter, it is the center frequency of the passband or the maximum point of rejection.

FILTER Q

The letter Q is an abbreviation for the *quality* of a circuit consisting of inductance (L), capacitance (C), and resistance (R). These are the components of which a filter is made. The Q of such a circuit does not include capacitance, since its quality is so high that capacitance is removed from consideration. The quality of a filter is directly proportion to the inductive reactance and inversely proportional to the resistance and can be expressed by

$$Q = \frac{2\pi f L}{R} = \frac{X_L}{R}$$

where f is the frequency in hertz, $\pi = 6.28$, L is the inductance in henrys, R is the resistance in ohms, and X_L is the inductive reactance in ohms. A high value of Q is desirable. The Q of a filter can be lowered by connecting a resistor in series or parallel with it.

An LC (coil-capacitor) filter is actually a tuned circuit. As such, it has a resonant frequency—that is, a frequency or a group of frequencies at which it produces the most voltage across it. Its output is independently adjustable and is determined by an internal variable-ratio oscillator. This ratio may be varied over a 300-to-1 range. External triggers may be used in place of the internal oscillator.

High-Pass and Low-Pass Filters

The *high-pass filter* favors treble tones, weakening midrange and low frequencies present in any source connected to its input. The cutoff frequency where the lows start to be attenuated may be adjustable with a possible range of more than 30 to 1.

The low-pass filter works in a manner opposite that of the high-pass filter. It passes low tones and weakens high frequencies present in any audio sound source connected to the input. The cutoff frequency may be adjustable, possibly in a range of more than 30 to 1.

High-pass and low-pass filters may be combined to produce band-pass or band-rejection filters.

Filter Control

Filter control determines the amount and type of variation in timbre. Its function is analogous to oscillator control. Variation in filter control encompass these results:

1. Envelope selection produces slow or rapid sweeps up or down in timbre.
2. Oscillator selection produces repeating changes in timbre.
3. Sampler selection produces sequential or random changes in timbre.
4. The filter control provides continuously variable tracking. Timbre can become duller or brighter or remain constant with increasing pitch.
5. The filter can be controlled by the lowest or highest key depressed. This permits the control of pitch with one key and timbre with a second key.
6. The filter control may have a provision for patching in an external signal for controlling timbre.
7. These effects may be combined to produce more complex variations.

The Sampler

The *sampler* produces a waveform for generating a random or ordered sequence of tones. A variable ratio oscillator is provided in the sampler for determining the sampling speed and to trigger the sampler's associated envelope generator.

Patch Panel

A synthesizer may be a relatively simple and inexpensive device equipped with few functions. The composer can upgrade in two ways: by obtaining a more sophisticated unit equipped with wanted features or by using the patch panel on the synthesizer. The *patch panel* permits studio expansion by adding keyboards and sequencers. Generally, standard-size phone jacks and plugs are used.

THE ARPEGGIATOR

If more than one note is being played simultaneously, an *arpeggiator* will cause the resulting tones to be played in sequence—that is, one at a time. Thus, an arpeggiator permits the tones of a chord to sound as successive tones and not simultaneously. The name of the device is based on the word arpeggio.

When in the arpeggio mode, if one key is missed, a major chord with that key being dominant is played. Thus,

$$C = C \text{ major cord}$$

$$D = D \text{ major chord}$$

$$A\# = A\# \text{ major chord}$$

Waveform Patches

In the early days of synthesizers, a number of its functions were handled by separate components. These were connected by cables known as *patches*, containing wires that were used for the transfer of electrical signals. When these separate components were finally combined into a single integrated unit, they no longer required patches. Modern synthesizers can be connected to other synthesizers, to machines that supply tempo, or to remote-control devices, and so the word *patches* has been retained. At one time patching simply meant interconnecting various components of a synthesizer, but the meaning of the word has now been extended to include an effect that is produced. Thus, a specific kind of beat is patched into a musical composition. Although this may have been done by a separate drum unit that required cabling between components, the same term is used when a synthesizer does not require external help.

Algorithms

Sometimes the word algorithm is used instead of or as a supplement to the word patch, particularly when components do not exist as discrete units, but rather are dedicated units. An algorithm could be the path followed by control voltages from one operating segment to another inside a synthesizer, as, for example, from an oscillator to a waveform filter to an amplifier.

The use of this technique to produce different voices results in a synthesizer equipped with a number of operators. These can be arranged in a number of different ways; their signal paths are called *algorithms*. An algorithm can be defined as a process that leads to a desired output from a given input. An algorithm is also defined as any mechanical or recurring procedure.

The block diagram in Figure 8-16 shows the technique used for producing a sound resembling that of a specific instrument. The sound begins with a VCO and this is the start of the sound source. The pitch (frequency) is controlled by a keyboard. The voltage output of the oscillator is fed into a VCF, with its output supplying the sound of a selected instrument.

The block diagram in Figure 8-16 is sometimes referred to as a patch and is the interconnection of a number of components. If the desired output is that of a guitar, the arrangement in the diagram can be called a guitar patch.

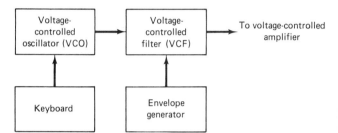

Figure 8-16 Patching of envelope generator to a voltage-controlled filter.

A patch can be complete or partial. The patch in Figure 8-16 is partial, since it does not include other components such as a VCA.

The patch in the figure is that of two VCOs whose outputs are being fed simultaneously into a filter. The VCOs can be set to the same frequency or not, depending on the music wanted by a performer. In some instances a noise signal may be wanted, and in that case the patch can consist of a VCO and a noise generator, with both connected to the input of a VCF. The VCO and the noise generator can be controlled independently.

Patches can have a number of variations, and the way in which circuitry is set up in a synthesizer is determined by its designer.

SIGNAL PATHS VERSUS CONTROL PATHS

Unlike circuit diagrams, in which electronic symbols are used, a block diagram consists of rectangles that represent complete circuits. These rectangles are identified by circuit names printed within their outlines. They are connected by vertical and horizontal lines, with the horizontal lines representing paths taken by signal voltages and the vertical lines, paths of control voltages. It is possible to have a patch using blocks and horizontal lines only. This patch can produce sound, but the sound is not controlled. Control can be exercised in a number of different ways. It can be done by a switch, by a variable element such as a potentiometer, or by a keyboard. The envelope generator produces a waveform for shaping pitch and loudness. It can be triggered by the internal oscillator, by a manual trigger, by the keyboard, or by other oscillators.

The timbre of a tone is set by the envelope of its waveform. The output of this generator can be directed to the VCA and the VCF. As previously indicated, the four elements of the envelope consist of attack, decay, sustain, and release. When the output of the envelope generator is connected to the VCF, the result is continuing control over the timbre.

GROWTH AND DECAY CHARACTERISTICS OF INSTRUMENTS

Figure 8-17 shows the growth and decay characteristics of a variety of string instruments. These are generalized waveforms and are not indicative of any specific type. Part (a) is that of a plucked string, (b) is a struck string, and (c) is a bowed string. All three are characterized by having a sharp attack time. This indicates that the voltage of the controlled circuit has jumped to a maximum amount as soon as the triggering signal was received from the keyboard.

There are two other controls that are involved: these are the *decay* and *sustain* controls. Decay determines how long the waveform takes to return to zero from its peak; sustain determines how long it will stay at its peak value before

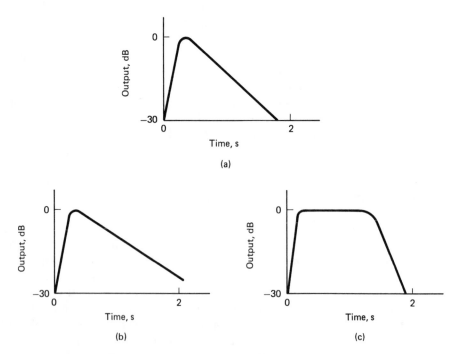

Figure 8-17 Growth and decay of sound. (a) Plucked string; (b) struck string; (c) bowed string.

beginning to decay. The drawings in (a) and (b) indicate very short sustain times; that in (c) has quite a long sustain time. Although the formation of a wave consists of four segments—attack, decay, sustain, and release—in some instances the time duration of one of these factors is so short that some waveform characteristics merge.

Figure 8-18 shows the waveform characteristics of three specific instruments: the piano (a), the guitar (b), and the organ (c).

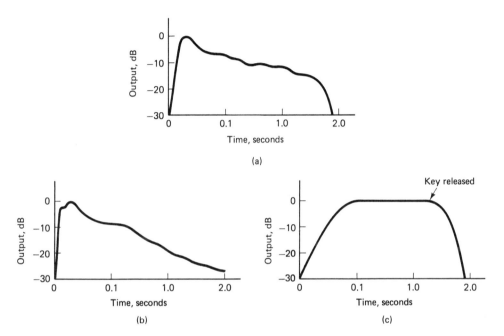

Figure 8-18 Waveform characteristics of (a) piano; (b) guitar; (c) organ.

Musical Instrument Envelopes

A note produced by a musical instrument will have varying amounts of loudness during its duration. This loudness can be represented graphically by a series of vertical lines, with each line indicating the amount of loudness at a particular moment of time. If all the tops of the vertical lines are connected across their tops, the result is a line known as an *envelope*. If this tone is increased in loudness, its envelope will remain the same. It will keep its identity; the only difference will be the increase in sound level.

The envelope is the distinguishing characteristic of an instrument. Thus, the envelopes of a trumpet and saxophone will be quite different. If we can duplicate the envelope of a trumpet electronically, it will sound like a trumpet. The same is true for all other envelopes.

SAMPLING

The conversion of an analog waveform to its digital equivalent is done by a process known as *sampling*. An analog waveform can be represented by a large number of vertical lines extending from the base of the wave to the envelope of that wave. Actually, the envelope is created by the instantaneous values of the wave—that is, its voltage value at any particular instant. Each of these instantaneous values or a number of them is known as a sample. A sample, then, is the digital representation of a waveform.

A synthesizer is equipped with sampling circuitry and is capable of accepting any waveform and supplying digital samples of that wave. The circuit that does the sampling is referred to as an A/D converter. The digits are then stored in memory and are available as required.

A digital sample can be played back by using a D/A converter. However, the played-back sample must be handled at the same speed at which the digitization is done. For this reason a number of samples, at different speeds, must be taken.

The advantage of sampling rather than synthesizing sound is that the sounds supplied by the synthesizer are more like instrumental sound.

Synthesizer versus Tone Generator

A synthesizer can be a complete, discrete unit, not dependent on outside components, or it can be made up of various types of add-on devices, such as a keyboard and a tone generator. A *tone generator* is controlled by an external keyboard, whereas a synthesizer is a tone generator plus its own keyboard. The two units, the synthesizer and the tone generator, may have the same number of features, or they may be somewhat different, depending on their design.

Dedicated versus Integrated Electronic Devices

Electronic components used in a studio can be dedicated or integrated. A *dedicated* unit is an electronic component that has a single function. An electronic flanger, for example, is an electronic component that does flanging and nothing else. A synthesizer may produce not only a variety of sounds but a number of sound effects as well. As such, it may house the equivalent of two or more components, and as such it is looked on as *integrated*.

A dedicated unit is desirable for updating a recording studio, adding new functions one at a time. An integrated unit supplies a number of functions.

Both types have their advantages and disadvantages. A dedicated unit can be less costly and may take up less room. However, using a number of dedicated components may call for extensive cabling. In the long run, using a number of dedicated components will require more room and will cost more. The great advantage of dedicated components is that replacement or updating means disposing of just a single component and not a number of them just because they are housed in one cabinet.

THE INTEGRATED SYNTHESIZER

Integrated and dedicated synthesizers each have their disadvantages. The operating panel of the integrated synthesizer can be quite complex. The dedicated synthesizer requires extensive cabling to its add-on components.

Not all integrated synthesizers are alike and some have more functions than others, which often are accompanied by a higher cost. In terms of functions an integrated synthesizer could contain:

1. *Noise Generator*. This is a sound source for producing wind, thunder, and percussive effects. It can generate white noise and can be used to produce random triggers and pitches when used in combination with other synthesizer functions.

2. *Master Oscillator*. This oscillator is a wide range, manually tunable oscillator. It produces a pitch proportional to the setting of its range switch and tuning control. It may contain size overlapping ranges to provide coverage in both the audio and subaudio spectrums—from 0.01 hz to 18,000 hz. The waveform output can be sine, triangle or square.

3. *Voltage Controlled Oscillators*. The synthesizer may contain two voltage oscillators (VCOs). Each oscillator has two control inputs with attenuators. The master oscillator, sampler, envelope generator and any optional keyboard can be used at these inputs to produce variations in pitch.

    The master oscillator can also be used to produce various types of vibratos of different speeds. The sampler (described later) can generate a series of ordered or random pitches. The envelope generator can produce sweeping pitches. The keyboard can supply discrete pitches of fixed intervals. The pitch of the VCOs can also be adjusted manually.

4. *Reverb*. Reverb is used to delay any audio signal connected to its input. In the process it adds body and depth to the original signal. The reverb may contain a depth control for proportioning the amount of reverberated and unreverberated signal and an output attenuator for adjusting the volume. The reverb can also be used as a waveform inverter.

5. *Electronic Switch*. This alternates two audio sound sources to a single output. The duration of each source at the output is independently adjustable and is determined by an internal variable ratio oscillator. This ratio may be varied over a 300 to 1 range. External triggers can be used in place of the internal oscillator.

6. *High-Pass Filter*. This filter passes high frequencies and attenuates lows that

are present in any audio source connected to its input. The cutoff frequency (where the lows start to be attenuated) is manually adjustable over a 30 to 1 range.

7. *Low-Pass Filter.* The low-pass filter passes low frequencies and weakens the highs present in any audio sound source connected to its input. The cutoff frequency is adjustable over a 30 to 1 range. The high-pass and low-pass filters can be combined to produce band-pass or band-reject filters.

8. *Microphone Amplifier.* The mic amplifier permits the introduction of external sound sources through the use of microphones and electric pickups.

9. *Modulators.* An integrated synthesizer can have two modulators with each capable of producing ring modulation. This is a circuit for mixing signals. It is so-called since it has four solid-state diodes connected in a circle. The signal input is to one apex of this arrangement; the output is taken from another apex.

   The diode circuit is passive and does not produce signal gain. The ring modulators can be used in combination with the envelope generators to supply amplitude modulation (AM).

10. *Stereo Mixers.* The synthesizer may have a pair of stereo mixers with each supplied with three inputs and equipped with volume attenuators. Through the use of pan controls the two mixers can be combined into a single mixer with six inputs.

11. *Pan Controls.* These permit any portion of the left mixer to be added to the right, or any portion of the right mixer to be added to the left; hence, they can be used to produce crossing effects.

12. *Envelope Generator.* This produces a waveform for shaping pitch and loudness. It can be triggered by the sampler's internal oscillator, an external oscillator, or a keyboard.

13. *Sampler.* The sampler produces a waveform for generating random or an ordered sequence of tones. A variable ratio oscillator is provided in the sampler for determining the sampling speed and to trigger the sampler's associated envelope generator.

14. *Patch Panel.* The patch panel provides inputs and outputs for every function contained in the synthesizer. This is particularly handy for the beginning electronic composer because it permits him to concentrate on one function at a time. The patch panel also permits expansion by adding keyboards and sequencers.

There are many types of integrated synthesizers, some with more features than those listed here, some with fewer. Some units are equipped with keyboards, some have a large number of voices, possibly in excess of 100, while some are equipped with MIDI input and output ports.

Voices

The word *voice* (Table 8-1) is used instead of the customary designation of an instrument by its name, such as hi-hat, violin, or flute, since the envelope can be modified electronically. The basic instrument sound is retained, but it is changed somewhat by the composer.

TABLE 8-1 PARTIAL LISTING OF VOICES

Bass Drum 1	Tom 1	Conga H mute
Bass Drum 2	Tom 2	Conga H open
Bass Drum 3	Tom 3	Conga L
Bass Drum 4	Tom 4	Bongo H
Bass Drum 5	Tom 5	Bongo L
Snare 1	Tom 6	Whistle
Snare 2	Tom 7	Cuica
Snare 3	Tom 8	Agogo H
Snare 4	Hi-Hat Closed	Agogo L
Snare 5	Hi-Hat Open	Tambourine
Rimshot	Crash	Shaker
	Ride (cup)	Bass 1 (thumb)
	Ride (edge)	Bass 2 (pull)
	Claps	DX Orchestra
	Cowbell	DX Marimba
	Timbale H	
	Timbale L	

MIDI

MIDI is an acronym for musical instrument digital interface. It is a useful arrangement for interconnecting electronic musical instruments or synthesizers. In effect, it is used to enable a single performer to control the output from a single unit. One of the advantages of MIDI is that it has brought about the compatibility of synthesizers, musical instruments, and computers made by different manufacturers. All the components in the MIDI system are linked by cables. For example, an interconnection can join a sequencer, a rhythm unit, and a keyboard. The keyboard—the control center—can be on one synthesizer, the rhythm unit on another, and the sequencer as part of still another. The MIDI arrangement can also include a computer for the storage of various rhythms and tones, and these can be taken out of memory as required.

A MIDI setup can be simple or it can be quite complex. A basic arrangement would be interconnecting a pair of synthesizers, but with only one equipped with a keyboard. The controlled synthesizer could supply drum set sound.

A MIDI is a piano keyboard device that has two basic characteristics: (1) It is a computer, and (2) it is an interfacing device—that is, it connects various electronic musical instruments and computers to each other by means of a cable.

The MIDI can control two or more synthesizers from a single keyboard. At the same time, it can also control an electronic drum set and the keyboards of the synthesizers. It is equipped with memory and can supply automated mixdown. It can also be used to permit one musical instrument, such as an electric guitar, to control a MIDI signal which, in turn, controls a synthesizer. That synthesizer can be set to its piano or possibly its accordion mode. Playing the guitar will then result in a sound like a piano or accordion.

The MIDI, when connected to a number of synthesizers, can produce the equivalent of small orchestra sound. The MIDI will synchronize the synthesizers and will also control an electronic drum set.

THE PERSONAL COMPUTER FOR MUSIC APPLICATIONS

Personal computers (PCs) are components that can not only be used for business but also as an aid for composers and musicians. One unit, shown in Figure 8-19, has a foldover screen (called a timbre door) that covers the keyboard when the unit is not in use.

The keyboard has standard IBM characters as well as music symbols that correspond to ROM-based music fonts, eliminating the need for special control codes and reference charts. Adding to its music applications design, there are two front-panel sliders that can be programmed for pitch bend, volume, and tempo. An additional picture-tube display can be had through the use of an external CRT (cathode-ray tube) jack. There is also a connection for an external monitor and a parallel printer port.

The PC is very good at keeping time. A dedicated internal music timer is provided to enhance the accuracy of music software. With SMPTE time-code in and out terminals, the computer is equipped for professional film and video work. The PC has connections for all standard computer accessories, including an external monitor, printer, modem, and mouse.

As in the case of other studio components, there are variations in music computers. The one shown in the illustration is MS-DOS compatible, includes MIDI and SMPTE connections, and is equipped with ROM- (read-only memory) based music fonts as well as full compatibility with MS-DOS-based programs for business and personal applications. This particular PC has 1 megabyte of internal memory. An optional 1.5-megabyte expansion board is available, which will bring the total memory up to 2.5 megabytes.

Some music computers feature an on-board 20-megabyte hard disk, whereas another will have two 3.5-in. 720K floppy disks. Both types are designed with extensive MIDI hardware, including 11 MIDI jacks (two in, one through, and

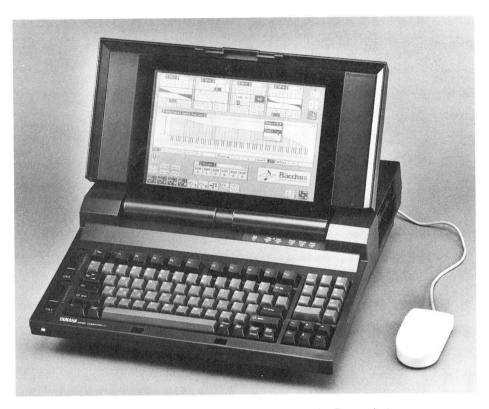

Figure 8-19   Music computer (Courtesy Yamaha Corporation).

eight out) and SMPTE in and out. (SMPTE is a timing code standardized by the Society of Motion Picture and Television Engineers.)

To emphasize the musical applications of this computer, there is additional hardware intended for music applications. For example, to give software developers even more flexibility, an additional system timer has been added. This consists of a chip that is not used by the main computer but is dedicated for music applications. A second chip assists in managing large amounts of memory. Known as a digital-music-application (DMA) controller, it permits direct memory-to-memory transfers.

CABLE CONNECTORS

MIDI-equipped components have five-pin DIN (Deutsche Industrie Norm, or German Industry Standards) sockets (jacks), more often referred to as ports, which are used to accommodate corresponding plugs. The pins of the plugs are arranged in a semicircle facing a keyway (Figure 8-20). The pins and the keyway are arranged so that it is impossible to insert a MIDI plug incorrectly into a port. In some instances, a five-pin XLR connector is used instead. The DIN and XLR connectors are not interchangeable.

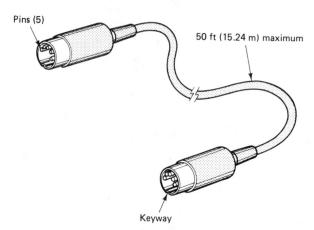

Pins (5)

50 ft (15.24 m) maximum

Keyway

Figure 8-20 MIDI cable with plugs at both ends.

Both DIN and XLR have five-pin plugs, although only three of the pins are used. The spare pins are simply used to lessen the possibility of pin movement, which is especially important if a MIDI is part of equipment that is often moved.

The MIDI connecting cable should have a maximum length of 50 feet (15.24 meters). The cable consists of a wire pair covered with flexible shield braid. The shield braid is the ground, or common connection, and is connected to pin 2. The so-called hot leads are joined to pins 4 and 5. These leads do not carry audio signals but are used for control voltages. For the five-contact XLR connector, the number 1 pin is grounded.

Port Arrangements

MIDI ports are identified by function, of which there are three. These are IN, OUT, and THRU (Figure 8-21). The IN port receives operating signals from other equipment that is MIDI equipped; the OUT port is the exit point of such signals; and the THRU supplies a signal equivalent to the IN signal and sends it along to other components.

MIDI
IN

MIDI
OUT

MIDI
THRU

Figure 8-21 MIDI ports.

MIDI SYSTEM ARRANGEMENTS

There are a number of MIDI system arrangements, including a MIDI-equipped piano keyboard connected to a multieffect processor; a pair of synthesizers that have MIDI circuitry; and a synthesizer, a rhythm unit, and a sequencer, all of which are MIDI equipped.

Master-Slave MIDI

One of the simplest of the MIDI arrangements is that shown in Figure 8-22, in which one synthesizer uses its MIDI circuitry to control another synthesizer remotely. Known as a *master-slave setup*, the master unit has a single cable connected from its OUT port to the IN port of the slave unit. Playing the keyboard of the master synthesizer results in functioning of the slave, with both master and slave working at the same time. Each synthesizer produces its own sound, resulting in an instrumental ensemble effect. Note that the slave synthesizer in this example cannot be used to control the master synthesizer.

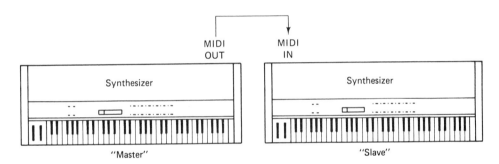

Figure 8-22 Master-slave MIDI arrangement.

With the master-slave arrangement, playing either a note or chord on the master will produce the same note or chord on the slave. If both the master and slave are touch-sensitive, the response on both will be the identical. In effect, the sounds will be superimposed.

Control data, not sound, are transmitted over the connecting cable from the master to the slave. Each synthesizer produces a different sound, and so the effect is that of a combo. Note that the master synthesizer is equipped with an output port; the slave synthesizer has an input port. Consequently, the data travel route is one-way.

Another MIDI master-slave arrangement, shown in Figure 8-23, uses a pair of connecting cables to join two MIDI-equipped components, a sequencer (sequence recorder) and a synthesizer. Both units are equipped with IN and OUT ports.

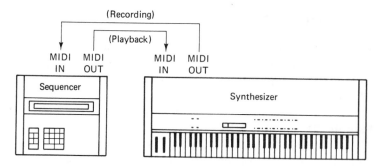

Figure 8-23   Master-slave MIDI using sequencer and synthesizer.

CHANNELS

The word *channel* indicates the transmission of specific control data. A MIDI setup may use from 1 to 16 channels, which is not intended to imply that separate wiring is required for each. The cable is only a three-wire type, and using separate wires for each channel would be impractical. Instead, coded instructions can be used for each required function. Whether a "slave" unit will respond to a specific instruction depends on the controlled unit. They are not all alike.

There are three modes of MIDI operation: omni, mono, and poly. *Omni* indicates the unit will respond to data sent over all channels; if *mono*, each voice in the controlled unit can be instructed to respond to a different channel. Finally, in the *poly* mode, a unit will obey data on its assigned channel.

A master unit does not mean an automatic increase in the number and type of the controlled voices. A controlled unit can supply only those voices with which it has been equipped, regardless of the instructions from the master. A monophonic synthesizer will play only one note at a time, even though its control unit is operating in the polyphonic mode.

MIDI Transmit and Receive

MIDI allows digital instruments to control each other in a variety of configurations. Data can be transmitted on any one of the standard MIDI channels to play a connected MIDI instrument such as a multi-timbral FM tone generator. It facilitates blending a digital orchestra with that of an electronic piano or the voices of a synthesizer.

THE ELECTRONIC PIANO

An electronic piano, MIDI equipped, can receive signals from an external keyboard, such as a digital programmable algorithm synthesizer, with this component used for playing the electronic piano. Alternatively a digital sequence recorder can send music that is too rapid and complex for human hands to play on the electronic piano.

An electronic piano has a built-in amplifier that can drive one or more speakers and a jack for headset input. Inserting the headset plug cuts off speaker sound so that this instrument can be used for no-speaker sound practice.

The voices that are used are based on samples of the sounds of actual instruments. Typically, five voices are available: reed, string, flute, brass, and percussion. Most often, the electronic piano supplies monophonic sound. Electronic pianos use different numbers of keys, but top-of-the-line units can have 88 and can behave like acoustic pianos, supplying pianissimo to fortissimo. Some electronic pianos are also equipped with a MIDI capability, enabling them to control a synthesizer or to be controlled by it.

MIDI LOCAL ON and OFF

Normally, an electronic piano is set to LOCAL ON, which means that its keyboard will play its internal voices. When set to LOCAL OFF, its keyboard controls only external MIDI devices. The electronic piano's internal voices will not be heard, though they may still be played by an external keyboard or sequencer.

MIDI Program Change

Pressing a voice key on the electronic piano also causes programs to change on connected MIDI devices. This function can be canceled so that MIDI program change commands are neither transmitted to nor received by the electronic piano.

MIDI Control Change

Transmission or reception of MIDI-control change signals affecting the sustain and soft or key-hold pedals can be canceled on the electronic piano. This permits sustaining the piano voice, for example, while creating a staccato trumpet melody on the synthesizer.

Augmenting the MIDI System

There are a variety of ways in which a MIDI system can be used. The MIDI-equipped electronic piano can be accompanied by a preprogrammed orchestral backing, supplied by a multitimbral FM tone generator. First, up to eight separate tracks of orchestral parts are recorded into the digital sequence recorder using the keyboard of the electronic piano to play the FM tone generator and select its voice programs. Then the orchestral arrangement is played back, simultaneously using up to eight of the FM tone generator's voices, while adding the live performance on the electronic piano.

The FM tone generator can be heard through the electronic piano's internal speakers or through any high-fidelity system. A slightly different technique is to use the digital sequence recorder to control the electronic piano; for example, the left-hand part of a classical piano piece can be recorded and then played back while practicing the right-hand part.

Another arrangement can consist of a digital rhythm programmer being used to provide simple rhythm patterns or a complex arrangement of drum and percussion sounds for an entire song. The digital sound processor permits adding a variety of stereo reverberation effects to the sound of the electronic piano. This creates the ambience of a concert hall.

Using a Digital Rhythm Programmer

The diagram in Figure 8-24 shows a synthesizer connected to a digital rhythm programmer. Both units have in and out ports, so data transfer can move back or forth. In this setup, the synthesizer can provide the basic notes for the performance. The digital rhythm programmer is used to add realistic percussion sound from its own pattern and song memories. When using the synthesizer's internal sequencer, the tempo of the digital rhythm programmer can be controlled via MIDI, or the digital rhythm programmer can be used to control the tempo of the synthesizer's sequencer. The synthesizer's sequencer can also be used to play instruments on the digital rhythm programmer, adding variation to the programmer's drum and percussion patterns. The programmer's instruments can be used to trigger the synthesizer's internal voices to add pleasing textural differences to the rhythm patterns.

Augmenting the Synthesizer and Digital Rhythm Programmer

The components used in Figure 8-25 can form the basis for a more elaborate MIDI setup, one that uses a synthesizer, sequence recorder, digital rhythm programmer, and tone generator.

In this example, the keyboard of the synthesizer is used to record notes and MIDI events on the sequence recorder. The sequence recorder can be used to replay these notes and events on the digital rhythm programmer using this component as a percussion tone generator, the tone generator, and the synthesizer. Alternatively, the digital rhythm programmer's own patterns can be synchronized using the MIDI clock generated by the sequence recorder. The digital rhythm programmer's echo-back function allows notes on the tone generator to be re-transmitted from the sequence recorder, while the digital rhythm programmer merges these notes with its own data.

MIDI Setup Using a Junction Controller

Figure 8-26 shows some uses of MIDI in a still more advanced arrangement. All MIDI signals are routed through a junction controller for different purposes.

When recording, for example, MIDI event data can be programmed into a

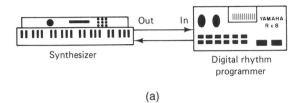

(a)

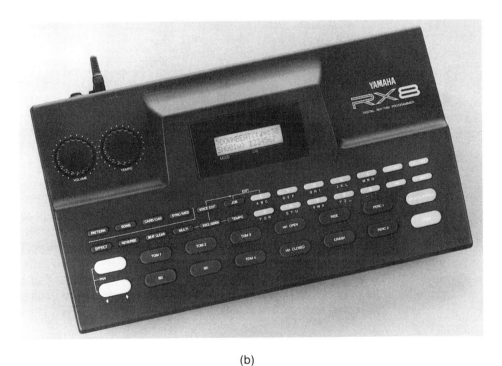

(b)

Figure 8-24   MIDI setup for synthesizer and digital rhythm programmer. Photo is that of the digital rhythm programmer (Courtesy Yamaha Corporation).

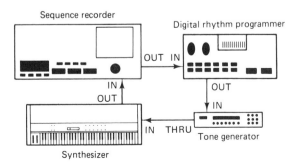

Figure 8-25   MIDI arrangement using synthesizer, sequence recorder, digital rhythm programmer, and tone generator.

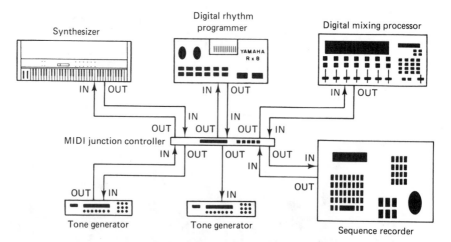

Figure 8-26   Circuit using MIDI junction controller.

sequence recorder from the synthesizer keyboard. The same keyboard using a different MIDI patch in the junction controller can then be used to program real-time dynamic information into the digital rhythm programmer's pattern memory.

On playback the sequence recorder controls the tempo of the programmer, using the sequence recorder's relative tempo function. The synthesizer together with the tone generator is controlled by the sequence recorder directly, as are the mixer settings on the digital mixing processor.

The sequence recorder can be used to store bulk data from the digital rhythm programmer , the tone generator, the synthesizer, and the digital mixing processor on its internal disk drive, using different patch settings in the MIDI junction controller.

Preprogrammed Orchestral Backing Plus Live Piano

The system shown in Figure 8-27 consists of a sequence recorder outputting into a multitimbral FM tone generator and then into an electronic piano. The tone generator supplies a preprogrammed orchestral backing. First, up to eight separate tracks of orchestral parts are recorded into the digital sequence recorder using the piano keyboard to play the tone generator and selecting its voice programs.

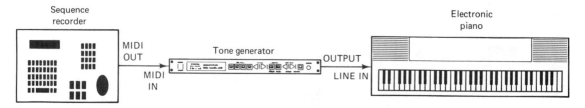

Figure 8-27   MIDI operation of an electronic piano.

Then the orchestral arrangement is played back, simultaneously using up to eight of the tone generator's voices, while live performance is added on the electronic piano.

Rhythm Accompaniment Plus Digital Reverberation Effects

It is possible to use the digital rhythm programmer to produce simple rhythm patterns or complex arrangements of authentic drum and percussion sounds for an entire song.

The digital sound processor (Figure 8-28) permits adding a variety of stereo reverberation effects to the electronic piano's sound, creating the ambience of a concert hall.

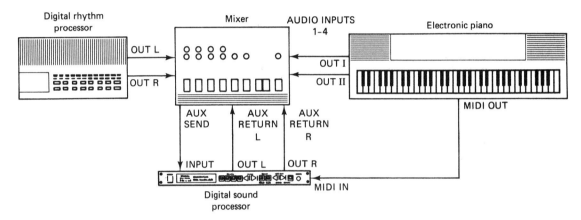

Figure 8-28 MIDI setup using rhythm accompaniment plus digital reverberation effects.

The processing unit also lets you add depth and vibrancy with a contemporary-sounding chorus or phasing program or wild modern effects with its pitch-change and delay programs.

With the power of MIDI, the electronic piano will automatically select programs on the processor each time you press a voice-select key, adding just the right effect to each of the piano's voices. The keyboard mixer lets you mix and pan the sounds of the piano and the digital rhythm programmer to create just the right balance, adjust the amount of processing, and even add an extra chorus effect for an enhanced, expanded sound.

SERIES VERSUS PARALLEL DATA TRANSMISSION

It would seem that using a sequencer to drive three or more synthesizers would be a good arrangement, but it does have its limitations. One possible setup, called a daisy chain, is shown in Figure 8-29. Any one or all of the synthesizers could

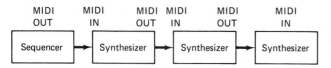

Figure 8-29 Series arrangement of a daisy chain.

be keyboard-equipped. This is a series technique, so-named because the data move from one synthesizer to the next in turn; this is done so rapidly that all the synthesizers appear to function simultaneously. Each of the synthesizers is equipped with IN and OUT ports, with the exception of the final synthesizer in the line.

While distortion is normally considered only in connection with music, it is also possible for it to afflict the data being transferred via cables from one synthesizer to the next. As a result, the final synthesizer—and/or possibly the one preceding it as well—may not play properly.

The solution is to use a MIDI THRU box; with this technique the data-travel paths are in parallel, as indicated in Figure 8-30. The sequencer furnishes the control data, which are then supplied to a MIDI THRU box. Four parallel paths are then supplied to each of the synthesizers.

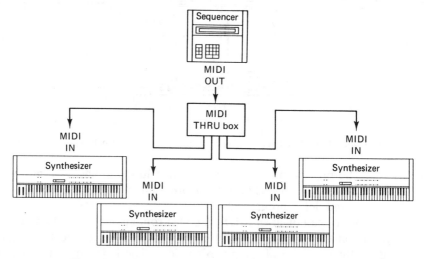

Figure 8-30 Parallel arrangement using a MIDI THRU box.

Keyboard to Drum Machine

An example of a series setup (Figure 8-31) involves the linkages between a MIDI-equipped keyboard and a MIDI-equipped drum machine. The drum voices can be controlled via the bottom octave of the keyboard. It is advantageous to use a pressure-sensitive keyboard for more positive control of the drum voices.

Since both the keyboard and the drum machine have IN and OUT ports, it

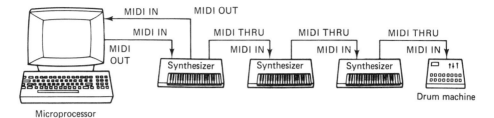

Figure 8-31   Series setup of a keyboard to a drum machine.

is possible to move the control in the opposite direction. This assumes the keyboard machine is sequencer- or arpeggiator-equipped. The advantage here is that the rhythm unit of the drum machine can then be made to determine the sequencer's or arpeggiator's playback rate.

MULTIEFFECT PROCESSOR APPLICATIONS

A *multieffect processor* can supply its effects to virtually any electric or acoustic instrument: synthesizer, guitar, piano, vibraphone, violin, or voice.

A MIDI keyboard can be used in conjunction with a multieffect processor. In this system the processor is connected to a MIDI keyboard and its outputs feed either directly powered speakers or a sound-reinforcement mixing console. The MIDI OUT terminal of the keyboard is connected to the MIDI IN terminal of the processor, permitting the automatic selection of different effects programs for specific voices selected at the keyboard. The processor is under the direct control of the keyboard player, rather than the mixing engineer, so the keyboard player can produce the desired effects for each voice or musical selection.

THE ELECTRIC GUITAR SYSTEM

The MIDI can be a discrete unit—that is, not an integrated unit combined with some other component, such as a synthesizer. The MIDI in this case can be a foot-controlled unit.

The processor may be able to supply effects such as flanging, phasing, echo, delay, and panning. Digital reverb can be an important part of any electric guitar system.

Its output jacks L (left) and R (right) are routed to separate guitar amplifiers. If the guitar system uses just a single amplifier, the output L and R is routed to the amplifier. In this way, both channels are sent to the single amplifier and reproduced monophonically. The guitar amplifier(s) are speaker-equipped. The line amplifier (amp) can be used to control the output sound level.

MICROPROCESSOR CONTROL

The techniques described in the preceding paragraphs have some disadvantages. They are manual, since they require the use of a number of manually controlled switching devices. Some use dials, but with these there is always an element of uncertainty, since it is not always possible to duplicate the settings precisely. Further, manual control takes time.

A simple way to make patch control faster, easier to use, and more precise is through the use of a microprocessor. This word is sometimes used synonymously (and incorrectly) with computer. The microprocessor (shown in Figure 8-31) contains an electronic memory. A program, or set of instructions, can be supplied to this memory, referred to as a ROM type. It has the advantage that its instructions cannot be changed or erased accidentally. The microprocessor can read its instructions from the ROM, but it cannot supply it with any new or different instructions.

9

Surround Sound

A two-channel stereo hi-fi system, no matter how excellent its sound quality, still acts upon a premise that is not true: the idea that sound comes from only one side—that is, the front of the listener. In reality, sound in any closed room, such as a concert hall, reaches the listener not only from the front, but also, in the form of reverberant sound, from the sides, rear, floor, and ceiling. This reflected sound (Figure 9-1) is what determines a room's or hall's acoustic behavior. Having traveled a longer distance than the direct frontal sound, the reflected sound reaches the listener a fraction of a second later. If this delay is large, it will give the impression of an echo; if small, it will help blend the instruments together, imparting a feeling of presence.

Surround sound takes these facts into account, providing rear sound through speakers. Some of the various arrangements for doing this are shown in Figure 9-2. The first of these is called the 4–0 system; although all the speakers are positioned in front of the listener they are positioned at different heights. In the 3–1 system, one speaker is placed behind the listener; three are in front. Still another technique is to put two speakers in front and two at the rear, a method referred to as the 2–2 system.

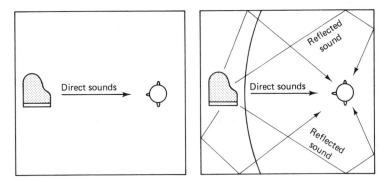

Figure 9-1 Even with a centrally located solo instrument, such as a piano, sound reaches the listener from all directions.

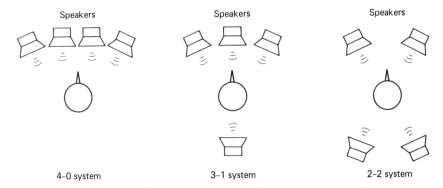

Figure 9-2 Some possible speaker arrangements for surround sound.

STEREO SOUND

Figure 9-3 shows the recording and reproduction processes of a two-channel stereo system. If this system is used to make a live recording, the mics are positioned to pick up sound to the left of center and to the right of center. Two channels are created, and these require separate amplification. In the reproduction process the listener, using a stereo system, is ultimately able to get the auditory impression of a spread orchestra—that is, the sound does not seem to come from a single sound source located somewhere on stage center.

   This stereo system does pick up reverberant sound—sound that is bounced off walls, floor, and ceiling, but this sound is incorporated with the two-channel recording, is indistinguishable from it, and is not reproduced for what it is—a separate sound entity. What is missing then is the beauty and richness of sound contributed by the acoustics of the music hall, an effect enjoyed by those who attend live concerts but lost by those who want the convenience of home listening.

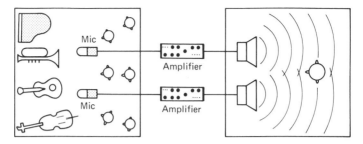

Figure 9-3   Stereo system produces spatial effect through the use of two sound channels and two speaker systems.

Sound Phase

With two-channel stereo, all sounds—direct, indirect, and of different phases—are heard from two front speakers. In a music hall sound from the front reaches the listener sooner than reflected sound. This time difference, also called *phase difference* (or simply phase) contributes immeasurably to the richness and beauty of sound.

EARLY SURROUND SOUND

During the 1970s various efforts were made to introduce sound that consisted of stereo, augmented by independent reverberant sound. Known as *four-channel sound* and also as *quadraphonic* sound, the three main proponents were Sansui with a system which it called SQ, CBS with a system it referred to as QS, and Discrete, promoted by RCA. None of these incompatible systems succeeded. Television sound was not adaptable, since it was monophonic.

In retrospect, quadraphonic sound of the 1970s generally attempted to optimize spatial reproduction of music around a full 360°. However, this proved to be beyond the capabilities of two-signal channels, except, perhaps, to a limited degree for a few people near the center of a regular speaker array.

Today, television sound is available in both mono and stereo, and many TV productions have been produced in surround sound. And it is now possible to transform even monophonic TV sound into simulated surround sound using sophisticated synthesis techniques.

Virtually all current films have a soundtrack recorded with the Dolby© Stereo or Ultra Stereo multichannel process. All videocassettes or compact discs of these movies include multichannel sound tracks, but these are not reproducible without some add-on modifications.

Present surround-sound techniques can add greater sonic realism not only to TV broadcasts but to all mono or stereo recorded material, whether these are vinyl LPs, digital audio tapes, compact discs, or audio cassettes.

Sound Channels

In stereo, to distinguish one channel from the other, left-channel sound is identified by L and right-channel, by R. In stereo sound transmission, the carrier is modulated with L and R signals, usually represented as L + R. For monophonic receivers, the L + R signals are combined to supply monophonic sound. For stereo reproduction two factors are required: One is a circuit for separating L and R signals; the other is a circuit to alert the receiver that stereo signals are being transmitted.

To achieve this latter objective when transmitting a stereo signal, the broadcast station transmits a subcarrier signal separated from the main carrier wave by 38 kHz together with a pilot frequency of 19 kHz. This latter signal is within the audio frequency range, but once the pilot signal has done its work, it must be eliminated to avoid high-frequency interference. This is done by using a filter that is poorly designed and is also capable of removing upper-frequency treble tones.

The Multiplex Decoder

A multiplex (MPX) decoder is a circuit in a stereo tuner or receiver that extracts left- and right-channel signals from an FM MPX broadcast. The advantage of multiplexing is that two or more channels can be transmitted on a single FM carrier. Actually, the R channel is transmitted in the usual manner, but the L channel must be encoded into the transmitted signal. In a monophonic receiver— that is, a receiver not equipped with a decoder—only the R channel information will be processed.

Stereo Signal Encoding

For stereo, a minimum of two mics are required. The first step is to use a pair of mics, identified as L and R, with each mic output fed into its own amplifier. The output of these amplifiers consists of the combined audio signals and are now called L + R. These signals can be modulated onto an RF (radio-frequency) carrier and transmitted. However, this is just a first step toward stereo.

The next move is the inclusion of a 38-kHz subcarrier. This carrier is produced by an oscillator, and it is combined with the L + R signal prior to modulation of the FM radio-frequency carrier. This carrier can be an FCC-assigned frequency between 88 and 108 mHz for the FM broadcast band.

Phase-Lock Loop Circuit

Although a stereo FM receiver could use a manual tuning circuit to lock onto the 19-kHz pilot signal, it is best done automatically and is accomplished by a phase-lock loop (PLL) circuit. This circuit permits the MPX demodulator to separate the stereo signals from the FM MPX signal.

The L and R signals are both audio and, as such, require separate audio amplifiers and separate speakers as well. For stereo, then, a minimum of two speakers is required. Stereo receivers are sometimes called FM MPX to indicate they are capable of supplying L and R audio signals. However, they are also equipped for supplying monophonic sound.

In an FM MPX receiver, the output of the demodulator is fed into an MPX decoder followed by an MPX noise filter. The L and R signals are fed into separate integrated amplifiers—that is, combined pre- and power amplifiers—followed by individual speaker systems. For mono pickup, combined L and R signals are fed into both L and R amplifiers and speakers.

SPATIAL DIRECTIONALITY

One of the unique features of perceived sound is that we can determine its approximate direction, more technically referred to as *spatial directionality*. It is this characteristic that lets us differentiate between the sound produced on a stereo playback system and the music of a live performance. There are other musical features that are important: accurate frequency response, dynamic range, the amount of freedom from distortion, sound reverberation, and so on. But it is sound directionality that supplies the necessary clue as to whether perceived sound is real or artificial.

Phantom Imaging

Also known as *virtual imaging, phantom imaging* is the ability of a pair of loudspeakers to create a phantom image between them, which sounds identical to what would be heard from a single loudspeaker at that point.

Multichannel Sound

In *multichannel sound*, three front channels, plus one channel in the rear reproduced by two or more loudspeakers, are used to create an encompassing sound environment. Regardless of where a person sits in a room, the three-channel front system stabilizes the image of sound to the front sound stage more accurately than any two-speaker system, thus adding unparalleled realism. The rear channel allows sound to be placed anywhere from front to rear.

The most important sound elements occur at front center stage. This is where we typically hear the soloist, either vocalist or lead instrumentalist. It is also where the greatest energy of most orchestras and other program material occurs. Unlike two-speaker stereo, multichannel sound has no "hole in the middle," no matter how wide the sound stage.

The front left and right loudspeakers each form a pairing with the front center, allowing sound to be reproduced anywhere to the left or right of center.

Finally, the rear loudspeakers open the sound and provide the spatial dimensionality of the concert hall or the theater.

The multichannel format also provides an expansive listening area, whereas two-speaker stereo limits the accuracy of sound imaging to a small area.

With conventional two-speaker stereo, only the listener located exactly between the two speakers perceives the correct stereo image location. A listener off-center will hear a shifted, or distorted, stereo image location, represented by the performer at the left in Figure 9-4.

Figure 9-4   Off-center listener hears shifted, or distorted, stereo image location (Courtesy Shure HTS).

With three-front-speaker, multichannel sound, it does not matter where the listener is seated. The image is heard in the same location by all listeners (Figure 9-5). This format's advantages can by realized only if each of the loudspeakers produces all directional sounds from as small a point as possible and directs them uniformly over the entire listening area while reducing reflections from nearby surfaces.

A single loudspeaker capable of this would be ideal but does not exist. Therefore, the drivers must be as close together as possible in order not to blur the image. Very low frequencies below 80 Hz, such as those generated by a subwoofer, produce wavelengths of 14 ft (4.3 m) or longer and are essentially nondirectional. Directional sounds can be considered as being at least twice those frequencies, or having wavelengths below 7 feet (2.1 meters). A midfrequency such as 1 kHz has a wavelength of only 13 in. (33 cm). These represent easily localized sounds if reflections from room surfaces near the loudspeaker or cabinet vibrations do not interfere with the primary sound source.

It is essential that the loudspeakers not be directional, since this would

Figure 9-5   With three-front-speaker, multichannel sound, the image is heard in same location by all listeners (Courtesy Shure HTS).

produce undesirable reflections from surrounding surfaces. Also, a highly directional loudspeaker severely limits the size of the desirable listening area. Most high-quality stereo speakers have a highly directional beam of sound at high frequencies, a broader beam at the middle frequencies and are much more omnidirectional at lower frequencies than they should be.

At the crossover between drivers the beam can fluctuate wildly. This does not really cause very noticeable problems in two-speaker stereo because the two speakers reproduce all the sounds from the front to the rear of the original sound field. In other words, the sound field is already audibly limited with only two-loudspeaker reproduction. In a multichannel setup, frontal sounds are reproduced at the front, and rear or ambient sounds from the rear.

When adding loudspeakers for a home audio/video entertainment system, give the highest priority to the center front speaker. This is the most important speaker in the system. The subwoofer is next for increasing realism and dynamic range. Then add the right and left front speakers to complete the front sound stage. Finally, add the surround speakers for the rear channel.

Location of Sound

The location of sound is essentially done by sensing the differences in loudness and in the arrival time of the sound at each ear. This information is further enhanced by the human brain's ability to sense a comb-filter effect (Figure 9-6), a series of notches or dips in the frequency response of the ear itself that affect all sounds that are heard.

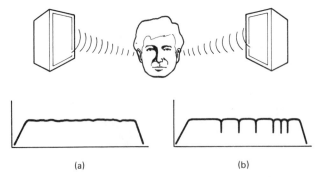

Figure 9-6 (a) When measured in an anechoic chamber, loudspeakers can supply a reasonably flat frequency response. (b) Our ears and brain, influenced by a physiological comb filter, perceive the sound as flat.

The comb filter is so called because a graph of the response looks like a comb. The notches shown in the waveform are caused by outer-ear resonance and diffraction effects—that is, by the interference of direct and reflected sound waves and partially by resonances within the ear canal itself.

The comb-filter effect is a natural phenomenon that gives us specific information about sound-source localization. We are never aware of the filter per se, but we can hear a difference when it is altered.

Stereo Imaging

There are various types of headphones, but they can commonly be categorized as circumaural or supra-aural. *Circumaural headphones* provide a sealed cushion around the ears; *supra-aural headphones* are open-air types, with the cushion resting on the ear. Both suppress or alter the ears' natural resonances so that the comb filter changes or disappears (Figure 9-7). Suddenly the realism and accurate sound localization that would normally be heard over a good set of monitor speakers in the studio are gone. However, if a headphone can retain, or even simulate the comb-filter effect, it will create an acceptably accurate stereo separation and localization. Localization is sometimes referred to as *stereo imaging*.

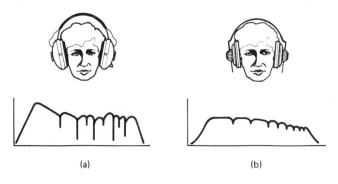

Figure 9-7 Sealed headphone under measurement conditions can produce better frequency response than loudspeakers. (a) In use, a tight seal emphasizes bass response and the comb filter's notches are altered. (b) Open-air headphones can also produce flat frequency response when measured, but in use the bass response is weakened. Ear pads damp the comb filter's notches.

Typical sound-recording techniques only aggravate the psychoacoustic problems caused by alteration of the comb filter. Nearly all stereo recordings are engineered to create natural separation when played back through monitor speakers. They have more than normal separation to account for acoustic mixing in the room during playback. However, headphones do not provide acoustic mixing or reverberant fields, so the separation tends to be very exaggerated. The combination of altering the comb-filter effect and the exaggerated left-to-right separation in recording makes the headphone listening experience unnatural.

Anechoic Response

Loudspeakers yield a reasonably flat frequency response when measured in an anechoic chamber. Our ears and brain, influenced by their inherent comb filter, perceive the sound as normal and as flat.

When measured, a sealed headphone may produce a similar or better frequency response than a high-quality speaker. However, when worn, the tight seal of headphones around the ears exaggerates the bass response, while the location, number and depth of the comb filter's notches are also severely altered, destroying the impression of normal, flat sound.

An open-air headphone may also produce a similarly measured flat frequency response. However, air escaping around the ears weakens the effective bass response, while the earpads damp the desirable comb-filter notches, again restricting perceived normal hearing.

THE PROBLEM OF REAR-CHANNEL SOUND

During a musical event both direct (dry) sound and reverberant (wet) sound are recorded simultaneously. During playback, the two sounds, in stereo, consist of both dry and wet types. The room in which playback takes place now adds its own share of reverberant sound. The played-back sound, however, has lost its 360° spatial characteristic, with the ears localizing the sound to the front left and right speakers.

Four-channel sound is different. It extracts reverberant sound that has been recorded together with the front sound and treats it as though it were a separate sound source. It then sends the reverberant sound through an amplifier and finally through a pair of rear-positioned speakers.

With a four-channel system, then, the effect is a duplication of the sound but also of the conditions under which that sound was produced. Ordinarily, the installation of a stereo sound system is done in a listening room with little attention given to acoustics. To get the full benefit of four-channel sound, however, the room in which sound playback takes place should be considered as part of that system.

Stereo Sound versus Four-Channel Sound

Stereo sound consists of a pair of sound signals, usually indicated as left sound and right sound. These two sound signals are supplied to two speakers that are spaced apart from each other. If headphones are used, one earpiece is for left sound and the other is for right. Stereo supplies the illusion that the sounds are from various locations, but these are all in front of the listener.

Quadraphonic sound, also simply called quadriphonic, is a system that more closely approximates listening conditions in a hall or auditorium. It requires a minimum of four speakers (although more than four are sometimes used) to supply the illusion that the sound surrounds the listener. Sometimes it gives the listener the feeling of being among the musicians.

Surround-Sound Encoding

During a recording process four channels of sound are available: stereo which encompasses right-front and left-front sound, and reverberant sound which includes rear left and rear right sound. Stereo recordings include both left- and right-rear sound. These sounds are time-displaced from front-channel sound. During playback what is heard is the simultaneous reproduction of front and rear sounds; that is, the rear sound is not distinguishable from the front sound. However good the stereo sound may be, it is not a true reproduction of the sound as it is produced in a hall, auditorium, or studio. To achieve four-channel reproduction, the rear left and right sound must be treated as though they are distinct sound sources, and this is exactly what they are. This means the rear sound must supply localization accuracy and adequate dynamic range. Only two speakers are used for stereo reproduction, and this capability is also applicable to speakers for rear-sound reproduction, that is, just two speakers should be required.

Another requirement for adequate four-channel sound is that all four reproducers (speakers) must be in balance, with no emphasis of any segment of the audio spectrum.

THE SURROUND-SOUND SYSTEM

Figure 9-8 is a diagram showing the arrangement of the components in a Home Theater System (The Shure HTS). The heart of the system is a surround-sound processor-decoder, which receives its sound input from a television receiver (video monitor) or any one of a number of different stereo sound sources, possibly supplied by a high-fidelity sound system or a dedicated component such as a CD player or turntable.

The output of the surround-sound processor-decoder is connected to six power amplifiers, with each having its own speaker system. While all these units are supplied so as to function with the surround-sound processor-decoder, this

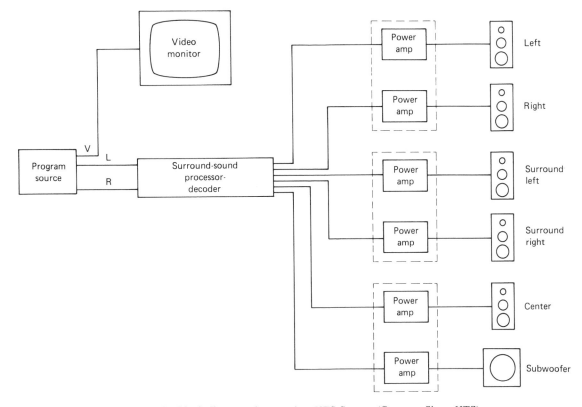

Figure 9-8 Block diagram of a complete HTS System (Courtesy Shure HTS).

unit can work with an existing high-fidelity sound system and accompanying monitor, laser disc player, VCR, and stereo cassette deck.

MATRIXING

Matrixing involves combining four signals into two channels in various phase and amplitude relationships. The assumption is that the original four signals can be recovered, but what is not specified is the extent of the recovery. While matrixed material requires a decoder, it can still be played through a usual two-channel stereo system; what will be heard will be stereo, not four-channel sound.

With a stereo-sound movie or a TV sports broadcast, this mode supplies the same kind of sound intensity that is present in a theater or stadium. The word *matrix* means adding and subtracting. A matrix network is an electronic circuit capable of adding signals to each other or subtracting one signal from another. There are various types of matrices; the objective of a matrix is to take apart signals that have been put together, recorded in two channels, adding some, subtracting some, then reassembling them into four-channel stereo. In matrix-type systems, the original four-channel sources are mixed or blended and then converted into a two-channel system, a conversion process called encoding. When the signal is to be played back, the encoded signals appear as four-channel sound via a process known as decoding.

Encoding is a means of packing four channels of information into two channels. While matrixed material requires a decoder for four-channel sound, it can be played through the usual stereo receiver, with the sound heard in two channels only. Figure 9-9 shows a representative matrix decoder.

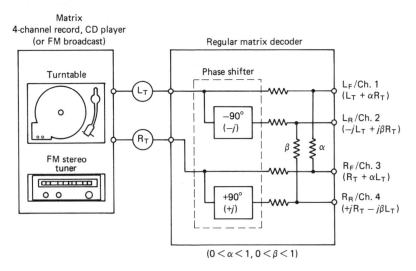

Figure 9-9 Possible arrangement of a matrix decoder. These steps in the functioning of a matrix decoder are listed in Table 9-1.

As shown in this illustration, signals L_T and R_T derived from a matrixed source pass through a phase shifter network in the matrix decoder and appear as four separate outputs. The drawing also shows that the portion of signal L_T is added to signal R_T to form the front-right signal R_T. The β portion of signal R_T, with phase led 90° ($+jR_T$), is added to signal L_T; with phase lagged 90° ($-jL_T$), it forms the rear left signal L_R, while the β portion of the $-jL_T$ signal is added to the $+jR_T$ signal to form the right-rear signal R_R (Table 9-1).

TABLE 9-1 STEPS IN THE FUNCTIONING
OF A MATRIX DECODER

L_F (front left, Ch. 1):	$L_T + \alpha R_T$
R_F (front right, Ch. 3):	$R_T + \alpha L_T$
L_R (rear left, Ch. 2):	$-jL_T + j\beta R_T$
R_R (rear right, Ch. 4):	$+jR_T - j\beta L_T$

L_T and R_T can be signals from an encoded phono record, compact disc, FM broadcast, or any other sound source that has been encoded.

The term $-j$ indicates that the phase of the signal has been lagged 90° with a phase shifter, while the term $+j$ denotes that the phase has been led by 90°.

Matrixing Problems

The fundamental problem of deriving four-speaker signals from two channels of sound is that the four reconstructed signals cannot be totally isolated and independent of each other. The mutual separation of the four, stereo sound and reverberant sound, cannot be infinite. Typically, each speaker signal has infinite separation from only one other speaker signal but only 3-dB separation from the remaining two speaker signals. This means that signals intended for one speaker can infiltrate two other speaker signals at only a 3 dB lower level.

This results in a significant deterioration of image localization, particularly for listeners positioned away from the center of the speaker array.

THE DOLBY© SURROUND PROCESS

Dolby© Surround is a 4-2-4 matrix process. In this process, four channels are encoded into two transmission channels. Information intended for the left and right channels remains intact. A center channel is encoded by applying equal levels of center information, in phase, to the left and right transmission channels. A surround channel is encoded by applying equal levels of surround information, in opposite phase, to the left and right transmission channels. The channels, also called L_T and R_T, become the standard left and right channels on consumer stereo software.

Two different approaches can be taken to decoding the directional information contained in L_T and R_T: a basic approach and one that is more sophisticated, which uses a technique known as *directional enhancement*. The basic approach is easy to implement but doesn't work very well. Imaging is very vague due to leakage of audio information between decoded channels.

Decoder with Directional Enhancement

In the case of a *decoder with directional enhancement*, the decoder looks for sound with strong directionality. If a left sound is sensed, for example, the decoder cancels the leakage of left sound into the center and surround channels. A small boost is given to the left channel to keep the total system power constant. As a result, the directionality of the signal is greatly improved.

Basic Problems

Applying four-channel sound to video tapes has raised three basic problems: level imbalance, phase mismatch and tape dropouts. The first of these, *level imbalance*, can be overcome by an input balance control on the surround-sound processor. *Phase mismatch* is caused by poor head alignment between the record and playback heads. This can result in enough out-of-phase information from dialog sibilants and other high-frequency sounds to cause a spitting effect in the surround channel. This problem is present regardless of the type of decoder used. It can be alleviated to some degree by the 7-kHz bandwidth and the time delay of the surround channel.

Tape dropouts are a fault that cannot be ascribed to any four-channel system and are more prevalent in low-cost tapes. The dropouts often occur more in one channel than the other, resulting in wandering acoustic images. In a standard stereo listening setup, the effect is noticeable. With a simple, nondirectionally enhanced decoder, the effect is usually less noticeable due to vague localization. With a properly designed directionally enhanced decoder, the localization is more precise than in stereo and so can make dropouts more noticeable.

With a hi-fi source and a decoder, a possible assumption would be perfect four-channel sound. This isn't always the case. The resulting presentation may be identical to theater sound, but that doesn't necessarily guarantee perfect sound.

CENTER-CHANNEL ENHANCEMENT

The surround-sound processor-decoder can be incorporated into an existing two-speaker stereo system. For cost reasons, the center speaker may not be included initially. While its desirability is justifiable, acceptable results can be obtained without it for centrally positioned listeners with narrow left-right speaker spacing.

If a center speaker smaller than the side speakers is used, it will probably have less bass response. Since bass frequencies are often mixed to the center position, activation of center enhancement could remove bass leakage from the side speakers and direct the energy to the center speaker with weak bass response. This could cause a drop in bass level and audible bass modulation. To avoid this problem, center-directed low-bass frequencies do not receive directional enhancement in the surround-sound processor-decoder. This does not introduce localization problems, since the low-bass frequencies do not contribute significantly to perceived directionality in the home environment.

Using a common subwoofer allows smaller speakers on the left and right. The surround-sound processor-decoder includes a subwoofer output with a fixed 12-dB/octave low-pass characteristic at 80 Hz. This balances well with many small, sealed-speaker systems with second-order roll-offs below about 65–100 Hz.

The requirement that the main speakers handle low-bass power may be reduced in a more elaborate system by the use of additional high- and low-pass filters to create higher-order crossover networks.

In the film theater, numerous small surround speakers are used to diffuse the rear sounds and provide even coverage. In the home, two surround speakers are more practical. The surround-sound processor-decoder uses its own acoustic space generator to diffuse the rear image and discourage localization at the closer surround speaker. The diffusing effect is similar to that obtained from the multiple-speaker array of a film theater. This does not reduce the ability of the directional enhancement circuitry to direct intended sounds solidly rearward.

Dolby© Surround processing circuitry enables this unit to become the center of a home sound system. Switching for conventional mono and stereo playback is obviously necessary.

With the requisite speakers and amplifiers in place, it is logical to provide a surround-sound synthesis capability for mono and conventional stereo signals, since these still make up the bulk of available sound-source material. The surround-sound processor-decoder provides a surround-sound-synthesis capability under user control.

An input-balance control and indicator are provided to adjust the balance of incoming L_T and R_T levels for optimum directional decoding. The input-level control and level indicator are used to calibrate the signal level through the decoding circuitry. Finally, the surround-sound processor-decoder includes a visual sound-directional display as an aid in system setup and in confirming proper system operation.

The Signal Path

The block diagram of the surround-sound processor-decoder in the Dolby© Surround mode is illustrated in Figure 9-10. L_T and R_T are first decoded into L', C', R', and S' by the basic decoding matrix block, which also includes the input level and balance controls. These four signals are modified by the directional enhancement circuitry into the enhanced signals L'', C'', R'', and S''. The first three of these signals are sent directly to the output-level controls and then to the output jacks.

The last of the four signals, S'', is first low-pass filtered sharply at 7 kHz and then passed through a reduced-effect Dolby© B-type noise-reduction decoder. The surround-time delay is provided by a wide dynamic range digital delay adjustable in precise 4-msec increments from 16 to 36 msecs. An acoustic space generator creates the image-diffusing left and right surround signals, which pass through the surround output level controls on their way to the output jacks.

The subwoofer output is an 80-Hz low-pass filtered version of C'. All six output-level controls are varied by the volume control, whereas a surround-level control varies only the level of the surrounds.

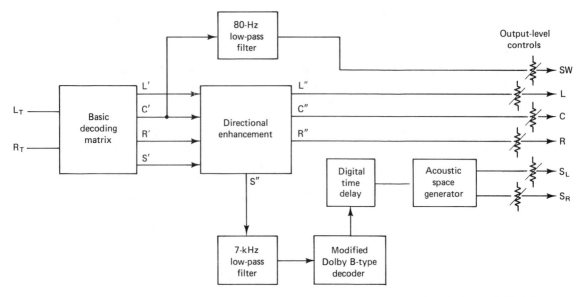

Figure 9-10 Block diagram of the surround-sound processor-decoder in the Dolby© Surround mode.

Total Surround Sound

The surround-sound systems of the 1970s were basically directed at phono records and recorded tape. They were not applicable to television sound whose monophonic approach raised problems. Compact disc and digital audio tape techniques were unknown. The modern trend, though, is toward the establishment of a complete in-home entertainment system, including all available sound-source systems, plus stereo sound television, both broadcast and in the form of video cassettes. As far as the latter are concerned, Dolby© Surround-encoded sound tracks are now widely available to the home market. This process uses a 4-2-4 matrix-based surround-sound encoding method.

The surround-sound processor-decoder decodes the information through its directional-enhancement circuitry, which stabilizes the localization of dialog and sound effects over a wide listening area, its wide-dynamic-range surround delay, and its surround image-diffusing circuitry.

SPEAKERS AND POWER AMPLIFIERS IN A SURROUND-SOUND SYSTEM

If an in-home entertainment system is to approach the capability of a modern motion picture theater, the loudspeakers and amplifiers must be capable of producing adequate sound levels in the listening space. To complete the effect, the diameter of the television receiver should be at least 27 in., preferably larger. Ideally, it should be a projection set. Large surround sound and a small TV screen would appear to be incompatible.

The three front loudspeakers, left, center, and right, should ideally all be the same. However, the speakers need only have similar characteristics to maintain the proper front-stage imagery. The similar characteristics are necessary because signals that are panned between two channels need the two sources to be similar to have a solid image. Also, if the same sound is moved from one channel to another, the character of the sound should not change. These three speakers should also be wide range with moderate dispersion.

The optimum listening area is greatly diminished when only four speakers are used. As a listener (or a viewer) gets even a few feet off-center, the central image will move to the closest speaker due to the increased sound level and the precedence effect. With *precedence effect*, the apparent source of a sound coming from two sources will move to the location of the source whose sound arrives at the listener's ear first. This effect is not a particular problem when listening to stereo sound.

Precedence effect becomes more significant when both video and audio are being used, since the eyes and ears are then able to perceive two different locations for the source of the same sound. The result is an uncomfortable feeling.

The surround speakers need not match the performance of the three front speakers but should still be of relatively high quality. Surround-channel information is bandwidth-limited to 7 kHz, but low distortion and modest low-frequency performance greatly improves the quality of the presentation. These speakers should have high dispersion to distribute the surround information around the back of the listening space. More directional speakers need to be aimed at the viewer for a quality presentation, and the optimum listening area is diminished. Floor-standing speakers or speakers with tall stands are the easiest to integrate into a listening space.

The subwoofer should be capable of filling the listening space with bigger-than-life bass, in the 30-Hz to 50-Hz range. The subwoofer channel of the surround-sound processor-decoder is low-pass filtered at 80 Hz. If a subwoofer is not used, the full bandwidth signals are available at the left, right, and center channel outputs.

When matching power amplifiers to the speakers, there are some additional considerations. Most people will use stereo power amplifiers for the additional channels. The power requirements for the two speakers used with the stereo amplifier cannot be much different. If they are, then either one speaker will not have enough power or one will be in danger of being damaged due to excessive power. This is likely to be of concern if the same stereo power amplifier is used to drive a subwoofer and a center-channel speaker, where the power requirements can be much different. Volume controls on the power amplifiers are an added convenience.

Room Lighting

There are two other factors to be considered in connection with a complete home-entertainment system. These are lighting and power turn-on. Generally, subdued lighting is preferred, possibly controlled by a dimmer. Many people appreciate the effect created by a small light behind the screen. This light is practical, because it closes the iris of the eye a little and improves the apparent contrast of the picture. Switchable power strips to turn the system on from a single location are convenient.

Locating the Loudspeakers

Figure 9-11 shows the preferred speaker placements for a variety of room arrangements. The center-channel speaker is located as close to the center line of the TV screen as possible, usually over or under the screen. The left and right speakers are then spaced as far apart as possible without developing a hole between the center speaker and the side channels. This will be controlled by the dispersion of the particular speakers.

Crossfiring the speakers helps fill in the hole. For a four-speaker presentation, having the speakers closer together is likely to be preferred due to precedence effect. It is also desirable to have all three front speakers at the same height above the floor. This helps keep the front image well balanced, especially for special effects and signals that are panned between two channels.

The surround speakers can be located anywhere from even with the listeners to a little behind them. The time-delay setting on the surround-sound processor-decoder can be used to accommodate a wide range of speaker placements. The best presentation is obtained with the surround speakers behind the listener at ear level or a little above. Crossfiring these speakers will help compensate for dispersion deficiencies.

Generally, subwoofers can be positioned at any convenient place in the room. These speakers sometimes come supplied with positioning instructions by their manufacturers and should be used as guidelines.

The Basic Decoder

The basic approach uses the left and right transmission channels as its decoded left and right channels. The sum of both channels becomes the center channel, and the difference between left and right transmission channels becomes the decoded surround channel. This approach is fairly easy to implement but doesn't work very well. Imaging is very vague due to leakage of audio information between decoded channels.

Figure 9-12 is a graphical representation of the encoded direction of sound sources. The left, center, right, and surround points arranged at 90° intervals represent the decoded channels of a basic Dolby© Surround decoder. The position

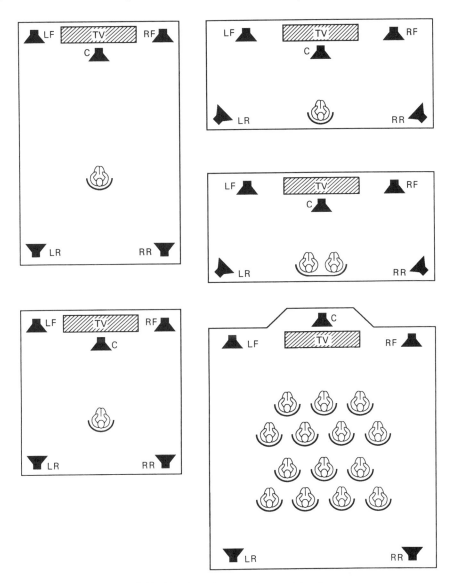

Figure 9-11   Speaker placement in a variety of rooms.

of an encoded sound on this circle represents the intended directionality of the sound. For example, a sound located at the left point on the encoding circle is intended to emerge from only the left speaker upon decoding. Unfortunately, this is not the case.

The angle between an encoding point on the circle and a decoded channel's location on the circle is related to the output of that channel. A 90° difference

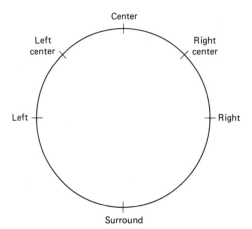

Figure 9-12   Encoding circle.

represents a cross talk signal 3 dB down in level, while a 180° difference results in no cross talk from that channel.

In the example of a left-encoded sound, the signal would appear at the left speaker at full level. The signal would also appear at center and surround loudspeakers 3 dB lower in level. The only channel without sound would be the right channel, since it is 180° from the encoded direction. A signal panned in between decoding points, such as a left-center signal, will appear in all four decoded channels, since no decoding point is exactly opposite the encoding point.

Directional Enhancement

In *directional enhancement*, the decoder looks for sounds with strong directionality. If a left sound is sensed, for example, the decoder cancels the leakage of left sound into the center and surround channels. A small boost is given to the left channel to keep the total system power constant. As a result, the directionality of the signal is improved.

CAPABILITIES AND LIMITATIONS OF DOLBY© SURROUND

The Dolby© Surround process, like any 4-2-4 matrix technique, has certain limitations in terms of its directional capabilities. When these limitations are exceeded, problems can occur in both theater and home environments. In theory, these problems shouldn't happen, because the monitoring setup will reveal them to the mixing engineers and they can be fixed at that point. In practice, however, some mistakes do get past the mixing stage and into the theater or home.

In the Dolby© Surround process, there are five basic encoding directions: *left, center, right,* and *surround,* and these are located on the encoding circle. The fifth, called the *interior,* is graphically represented by the inner portion within the encoding circle, as shown in Figure 9-13.

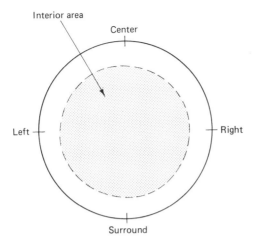

Figure 9-13 Interior area.

When a signal is encoded in the left, center, or right position, it is enhanced by the decoder and appears at only one speaker. The surround channel is similar, except that it is reproduced by the multiple rear speakers. Signals panned between left and center or between right and center are also enhanced by the decoder and emerge from only two speakers as phantom images. Signals present in both channels in equal amplitude and with a 90° phase shift are not enhanced and will appear at all speakers in equal amplitude. Signals within the interior—but not at the exact center—will also appear at all loudspeakers, but in varying amplitudes, depending on where they are panned.

This interior area is often used by filmmakers for environmental effects, music, and specific off-screen sound effects, because sounds mixed here surround the viewer, providing a good feeling of involvement. Sounds panned in this area do not interfere with the ability to enhance other sounds with strong directionality to their proper channel or 2-channel (phantom) position. For example, music and environmental sounds may be panned to various points in the interior area while dialogue and specific sound effects are simultaneously being panned to individual channels.

These signals will not fight one another; the interior mixed signals will remain as interior sounds, emerging from all speakers, whether or not directional enhancement of strongly directional effects is taking place, and the strongly directional sounds will emerge from only the intended one or two speakers, whether or not any interior sounds are present.

The Dolby© Surround process also allows enhancement of simultaneous sources placed directly opposite one another on the encoding circle. For example, center and surround or left and right sources can be simultaneously encoded and decoded.

Mixing problems generally occur when two or more primary sounds are mixed in adjacent positions on the encoding circle. For example, simultaneous left and center sounds will fight each other during the decoding process and cause audible shifting of the reproduced sound stage. To prevent this problem, one sound source can be panned toward an interior mix, or the two sounds can be staggered slightly in time.

Amplifier Power Output

The power output of the amplifiers is dependent on the power-handling capabilities of the various speakers in the surround-sound system. Sometimes an approximately 6-to-1 ratio is specified; that is, if the front-channel power is 70 watts/channel, about 12 watts/channel will be supplied to the rear speakers. This is for a modest system. A more desirable arrangement uses 100-W amplifiers working into an 8-Ω load. The reason for this arrangement is that the rear speakers carry only supplementary sound information, while the front speakers are intended for the original dry sound.

Digital Time Delay

Delay is used to make certain that sounds from the rear reach the ears at appropriate times, even though these speakers are much closer to the listener than those in the front. One millisecond of delay is roughly equivalent to 1 foot of distance, but even if all speakers were essentially equidistant from a listener, a delay of 20 ms would appear to be desirable as a starting point. If the front speakers are 8 feet farther away than the rear, add 8 ms of delay. If an echo becomes noticeable, reduce the delay. The final adjustment is a matter of personal taste.

Listeners seldom sit in the acoustic center of a loudspeaker array and are typically biased toward the surround loudspeakers. Hence a digital delay is used, in effect, to move the surround loudspeakers farther back into the listening space. Still another refinement of a decoder is its ability to select separate delay times to accommodate different listening environments, from large presentation rooms to small studio or home listening environments.

ACOUSTIC CONDITIONS FOR A HOME-THEATER SYSTEM

It isn't always possible to set up a listening and viewing room for optimum operating conditions in the home. Even a few modifications, though, may be better than none at all.

The ideal time for creating a room for a home-theater system is in the planning stages prior to building. If possible, make the room rectangular or, in an older home, try to select such a room. A good choice for the ratio of room dimensions is 1.0 to 1.6 to 2.5 (height to width to length). This ratio will distribute room acoustic response evenly at low to midfrequencies.

All closed spaces respond easily at certain frequencies, regardless of room shape or dimensions. These sound waves are called the *resonant frequencies* of the room's interior. The ratio just given for room dimensions is a good compromise for distributing these resonances evenly, since the room is excited by low frequencies from the audio playback system. The volume of the room is also important, since it will affect the actual room length, width, and height.

A good overall size is from 70 to 100 cubic meters. An excellent choice for the ceiling height is 3 meters, or about 10 feet. These dimensions can vary by 10% in either direction without significantly degrading performance.

Provide entrance to the room by one door (a maximum of two) of normal size. Multiple French door openings, although pleasing in appearance, are not the best choice for an entrance because they present large uncontrolled reflecting surfaces that tend to vibrate and rattle easily.

Also, multiple room sections with connecting passages that cannot be closed off from the home theater can contribute additional undesirable resonances at low frequencies and thus are not good acoustic design.

Avoid more than one or two windows (none is preferable), since they vibrate easily and extract sound-wave energy unevenly from the room. If windows are necessary, make the panes small sections and not large areas of plate glass. Cover the windows with fabric curtains that can be closed. Avoid concave window surfaces, such as bay windows, which tend to focus sound waves as light waves are focused by curved mirrors.

Place acoustical absorption on the wall where the video screen is located and on the side walls adjacent to the screen wall. The side-wall absorption should cover about 25% to 35% of the side walls, starting near the front-loudspeaker locations. The major part of the absorptive material should be near the front wall in the plane of the loudspeakers. Try to locate paintings or other large flat reflective surfaces on the back wall or toward the back of the side walls.

Acoustical absorption can be in the form of cloth-covered fiberglass panels or curtains. Random application of sound absorbing panel sections can be visually pleasing and better for the acoustics. Cloth-covered fiberglass tubes, 1 to 2 feet in diameter, positioned at the room corners help in the midfrequencies by damping the room's reverberant response. Tubes hanging from the ceiling in the corners are in good, out-of-the-way locations. Adding acoustical absorption to existing rooms whose dimensions or general layout are bad choices for good acoustics is a necessity to prevent degradation of a quality home-theater audio system.

Plan a dedicated wall for the picture screen. This wall should also be where the front loudspeakers are located. For large front-projection screens, position the screen (assuming a separate roll-type screen used with projection TV) near the ceiling, approximately 1 to 2 feet down, with the three front speakers in a line at the bottom edge of the screen.

Stand-mounted loudspeakers solve the problem most easily but flush-mounted speakers (in the screen wall) are also acceptable when the screen is on or very near the wall. The bottom edges of these speakers should be at least 26 in. from the floor. Above the screen near the ceiling is the worst choice for any of the front speakers: left, center or right.

The distance between the left and right speakers should be equal to the distance from the screen to the main viewing or listening position. This is a minimum distance; the width of a few feet more is even better for cinema programs (but do not place speakers at the room corners).

For large rear-projection TVs and monitors, place the center-channel speaker above the video monitor and the left and right loudspeakers spaced equally distant from the center at the height suggested before.

It has become somewhat stylish to have the loudspeakers out of view, but this will sometimes void the benefits of a quality audio system. Flush-mounting the speakers and attaching the supplied grilles works very well acoustically and still is unobtrusive. If designed correctly, however, loudspeakers, as in a movie theater, can effectively transfer the sound illusion to the home theater and will not affect the sound reproduction. The latter layout is preferred for the serious home-theater aficionado.

Large full-wall cabinet systems at the front of the room are generally poor locations for loudspeakers because of resonant cavity effects. Apart from creating poor acoustics, front-wall cabinet systems also require someone to walk in front of viewers to adjust the electronics in order to make a new program selection, if a remote control is not available.

A better choice is to place the wall cabinet system in the back (or side back) of the room to house the electronics, program material collection, and refreshments. Route the loudspeaker cables from the back wall to the front loudspeakers through the floor or along the bottom edge of the wall. At the back-wall location, the various cavities and different surfaces of the wall cabinet will help to diffuse the sound field in the "surround" loudspeaker's audio space and will not destroy the front sound-field image capabilities or defeat the overall function of a high-quality multichannel audio system.

Locate the subwoofer on the floor at the front wall or near the corner between the front and side walls. For multiple subwoofer systems, opposite locations in front and back can work well to provide complete coverage of the room.

Locate surround loudspeakers above and behind the normal listener positions. The best location for two speakers is generally on the back wall, separated in the same way as the front loudspeakers and typically 1 foot below the ceiling. A good minimum height for the bottoms of the speakers is 6 feet from the floor, or above the plane of the seated listeners' ears.

Consider four speakers as surround loudspeaker systems for larger rooms in a horseshoe-like array around the back and side walls. But do not position a side speaker next to the listening position.

For accurate low- and midfrequency reproduction it is desirable to reduce wall and ceiling vibrations induced by the home-theater loudspeakers, especially from the subwoofer. These secondary vibrations reduce the accuracy of the front sound images by reradiating the soundwave, which is then no longer correlated to the program. Most often the secondary vibrations also add annoying rattles. The most practical way to solve this problem and reduce organ-pipe-like action of sound waves in hollow walls is to make the wall heavy and without open cavities. To accomplish this consider the following added details for room walls, most easily done during new construction or during modification of the house structure.

For stud walls, place studs on 16-in. (or less) centers and pack the wall cavity between panels with fiberglass, even on an inside wall.

For wall surfaces built with gypsum board, use two layers of board on the room's inner walls. For example, use two $\frac{1}{2}$-in.-thick boards on top of each other with adhesive liberally applied between the boards. Plan the electrical fixtures accordingly.

For either wood- or metal-stud wall construction, use USG Acoustical Sealant on stud surfaces (between the stud and wallboard) and all cracks. This will reduce rattles considerably.

Plan for indirect lighting with "wall-washer"-type systems using dimmer

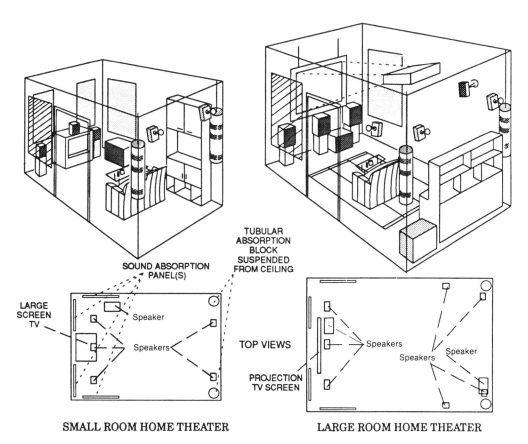

Figure 9-14 Small-room theater (left); large-room home theater (right) (Courtesy Shure HTS).

controls. As an additional touch, include lights that can be dimmed behind the screen and over the electronic equipment at the back of the room. Pick a quality dimmer unit whose manufacturer mentions that it is electrically quiet and that it suppresses radio-frequency emissions. Avoid fluorescent fixtures; use incandescent bulbs only. The dimmer should be a type designed to work with incandescent lights.

Provide at least one (maybe two, depending on the size and number of power amplifiers) 20-ampere branch power line leading directly from the main circuit box. Make sure the power line is a three-wire type: the black lead is the ''hot'' lead, the white lead is the neutral, and the third lead is a ground wire. The ground wire may be bare, but if insulated will be color-coded green. Use a separate, three-hole-type outlet (receptacle) for this power line branch. Connect the ground lead of the branch power line to a true ground such as a water pipe. Be sure to keep this branch as a dedicated power line; that is, do not use it as a power source for any other appliances. Do not try to make use of any other existing power line branch even if it is very convenient to do so. At the service box, use a 20-ampere time-delay fuse or an equivalent circuit breaker.

Figure 9-14 shows suggested arrangements for small-room and large-room home theaters.

MOTION PICTURE (MP) VERSUS HOME SURROUND SOUND

Surround sound in motion picture theaters supplies music, dialog, and panned effects using a three-channel front soundstage with surround effects and ambience coming from the sides and rear of the theaters. These four channels are encoded or matrixed into two-channel Dolby© and are decoded in the theaters by a professional Dolby© stereo MP decoder. The surround-sound effects are hidden in a two-channel stereo movie release and decoded in the theater by special Dolby© equipment.

These same two-channel MP masters are used for the production of stereo VHS, VHS hi-fi, and stereo laservision tapes and discs that can be purchased or rented in video stores. To take advantage of this technique in the home, a surround-sound receiver can be equipped with a Dolby© Pro Logic Surround capable of creating a multidimensional soundstage. As in a theater, an active center channel ensures that all members of the audience hear dialog coming from its on-screen source, regardless of seating positions.

If the surround-sound receiver is equipped with a front amplifier and a center amplifier, Dolby© Pro Logic Surround, Hall, and matrix effects can all be produced. To reproduce these effects, rear (surround) speakers and center speakers are required. The placement of these speakers is important. It is helpful to experiment with various positions and locations to find the best placement for optimum sound quality.

Dolby© Pro Logic Surround

When using video cassettes or video discs that carry the Dolby©, Stereo or Dolby© Surround trademark, it becomes possible to achieve the same kind of sound in the home that is heard in a theater. The main Dolby© Pro Logic Surround functions are sent through the center channel. Because of this, three modes must be available from the speaker used by the center channel. There are three possible modes.

Normal. When small speakers are used in the center channel, signals of 100 Hz or more are output and signals of 100 Hz or less are split between the front left (L) and right (R) channels.

Wide. When the front L and R channels use similar speakers for the center channels, the sound system operates throughout the full frequency range.

Phantom. When no center speaker is used, the center channel is split between the front L and R speakers, producing the same results as with a center speaker.

10

Recording and Recording Studios

BASIC ANALOG RECORDING

Recording can involve nothing more than a single mic and a cassette deck. This simple approach, used for home recording, involves a minimum amount of equipment and is characterized by a lack of versatility. Recording studios can also have limited equipment, with no attention paid to studio acoustics. At the other end of the scale, professional studios use a variety of sound-processing devices, producing tapes suitable for submission to record or compact disc manufacturers or radio and TV broadcasting stations. One of the big differences among recording studios is also the extent to which they utilize both analog and digital equipment and the sophistication and operating features of that equipment.

Devices for studios that are analog-oriented always include two or more mics followed by amplifiers. Faders may be used for control of signal amplitude. The audio signals are brought into a mixing console, and it is here that the sound engineer regulates, controls, reduces, or increases the signal. Various sound effects are also added. Individual instruments or groups of instruments are then recorded on a multitrack tape deck. These tracks can be mixed down or altered by the sound engineer until all that remains is a pair of tracks, one for left-channel sound and the other for right-channel sound. In some studios, recording and mixing is done in a single console instead of using a dedicated recorder and separate mixer.

BASIC DIGITAL RECORDING

Digital recording involves audio editing and a record and compact disc production system. A number of microphones are used, with their respective outputs brought into audio amplifiers preceded by faders. The signal is then brought into a mixer console. The input signal to the mixer need not be microphone-generated but can consist of one or more synthesizers, which may or may not be MIDI cable equipped. The mixer may also have MIDI circuitry for controlling the synthesizers, assuming these are equipped with ports for the acceptance of MIDI signals.

The output of the mixer is then supplied to an analog-to-digital (A/D) converter, following which the signal moves into a digital mixer. A digital reverberation device is also connected to this mixer. The mixer signal can be supplied to a VCR or to a tape deck and then to a digital audio editor and also a digital audio processor. This processor controls a digital cutting machine, following which the digital signal is stamped into a compact disc.

A tape machine is used in both analog and digital recording, but there is a difference in the results. An analog signal has a varying amplitude, with the signal ranging from zero to some peak value, depending on signal loudness. When the signal is very weak, it can extend into the noise floor of the tape. The noise floor is the signal produced by moving tape in the absence of a signal to be recorded. At the other extreme, a peak signal is capable of saturating the tape, resulting in sound distortion.

These two possible problems are eliminated in digital recording, because all the pulses representing binary numbers have the same amplitude. Consequently, the recording machine can be set well above the noise floor and substantially below the maximum peak recording level. The dynamic range of the tape is no longer important. For analog sound, the dynamic range is about 60 dB, but for digital recording it can be as much as 96 dB.

Another disadvantage of recording analog signals is that efforts are usually made to put as much signal onto the tape as possible. Since tape is tightly wound, it is possible for the signal to print through from one layer to the next, resulting in echo signals.

The digital recording process also eliminates rumble, wow, flutter, and intermodulation (IM). The frequency response is practically flat from 20 Hz to 20 kHz, and harmonic distortion is below 0.05%.

MICROPHONE TECHNIQUES

Recording single instruments does not always require a "cookbook" placement of the microphone. Recommendations made in this section are based on standard studio practice, and experimentation with placement can provide interesting sound textures.

Micing the Flute

To reproduce the distinctive sound of the *flute*, select a mic having a smooth, wide frequency-range response. To record flute sound without distortion, position the microphone at right angles to and about 1 foot away from the instrument, facing the performer's mouth. Place the mic slightly below the flute, as indicated in Figure 10-1, and looking upward at the flute so as not to be in a direct line with the air stream. Alternatively, place the mic midway between the mouth and the finger position to lessen any overblow effect. The problem with recording the flute is that it has a very delicate sound output and is unable to compete with instruments supplying a higher sound level. This is not only true when used with other instruments but also when the flute accompanies vocalists, such as a solo female singer.

Figure 10-1 Micing the flute (Courtesy AKG Acoustics, Inc.).

Recording the Flute

For single-mic pickup, position the mic right at the mouthpiece, but with this arrangement be sure to use a windscreen. An alternative arrangement is to have the mic about 6 in. to 12 in. from the embouchure of the instrument.

To emphasize flute sound, mic in as closely as possible, using breath sound as a guide. Move the mic in until you hear breath sound, and then back off slightly. Also try using a mic having a built-in breath-noise control.

The flute represents more of a recording challenge than many other instruments because the sound energy is projected both by the embouchure and by the first open fingerhole. In other words, the flute radiates sound as a dipole, projecting no frequency range, not even the lowest, all around. Some recordists use two mics. One is used to pick up the sound range up to 3 kHz. Place a mic along the flute player's line of sight, with the angle at 0°. Use the second mic for the range above 3 kHz to the right of the flautist, with the angle at 90°.

The working distance depends on the style of music played. For pop and jazz, relatively short distances may be desirable, from 1 to 2 in., as the amount of wind and breath noise rises with decreasing working distance. In this case the flautist should blow a little below the mic to avoid getting too much noise or overloading the mic.

Recording Reed and Brass Instruments

Reed instruments, such as the clarinet, saxophone, oboe, and the reed organ (harmonium), the harmonica, and brass instruments, such as the trumpet, trombone, french horn, and tuba, are often played solo or as part of an instrumental section in an orchestra. Position the mic near the bell of reed instruments that are so equipped to get the full tonal coloring. Cardioid microphones for solo instruments and two-way cardioids for reed sections give good results. In large orchestras where it is desirable to have separation of sound control between reed sections and brass sections directly behind them, a bidirectional microphone or out-of-phase cardioid pair, placed back to back, will give a full reed sound without brass pickup, because the dead (or null) side of the bidirectional, or figure-8 pattern is toward the brass. Brass-instrument soloists often play to the side of a cardioid microphone to reduce the high-frequency bite of the instrument. The reduced high-frequency response at 90° off-axis on a single-element cardioid may be useful in controlling the brilliance of brass instruments. In recording brass sections, cardioid microphones placed several feet from the section give a good section blending of instruments. Of course, if greater separation of instruments is necessary use close mic placement as required. This will provide excellent detail to the music, while more distant placement provides greater blending of instruments.

Micing the saxophone. One of the problems in micing musical instruments, as well as vocalists, is the pickup of sound not directly related to the audio that is wanted. Breathing noises, as in the case of vocalists, bowing sounds produced on the violin, and key noise produced on the saxophone may be desirable or not, depending on the listener. Some find these sounds add realism; others regard them as noise and a sonic intrusion.

During the time it is played, the *saxophone* produces key noise with the amount that is evident to a listener depending on location with respect to the instrument. It is also dependent on the working distance of the mic from the instrument. The closer the mic, the more possible it is to hear key functioning.

To include key noise, point the mic directly downward at the bell opening of the saxophone, and, if necessary reduce the working distance. To avoid key noise, direct the mic toward the front edge of the bell. It will be necessary to experiment to avoid wind noise. In either case, the working distance of the mic should be somewhere between 8 to 12 in., depending on the sensitivity of the mic and the effort being expended by the musician (Figure 10-2).

Figure 10-2 Micing the saxophone (Courtesy AKG Acoustics, Inc.).

The data supplied here are applicable to all eight types of saxophones. Only four are commonly used, however; these include the baritone, tenor, alto, and soprano saxophones. The saxophone is a reed instrument and uses just a single reed, and so from this point of view it may be said to be related to the clarinet. The instrument may have a flared bell or else simply a conical tube; in that sense it could be considered more closely related to the oboe.

Micing the trombone. The *trombone* is a brass wind instrument having a cupped mouthpiece, with the instrument ending in a bell. The mic can be mounted on an adjustable stand, with the mic facing the center of the bell. The trombone has considerable sound output and so the working distance of the mic must be found experimentally. A customary distance is 12 in. (Figure 10-3).

Micing the clarinet. Unlike the trombone, whose sound comes from the bell, almost all sound from the *clarinet* exits from the finger holes, although there is sound from its small bell. The mic, mounted on an adjustable stand, is aimed halfway along the finger holes (Figure 10-4).

Micing the harmonica. The *harmonica*, also known as the western mouth organ or simply the mouth organ, is a reed instrument with relatively low sound output when compared to other reed instruments. Figure 10-5 shows the micing technique. The objective is to bring the mic as closely as possible to the instrument. It may be helpful to use a mic characterized by proximity effect in order to put more emphasis on the bass tones. Usually the performer holds the harmonica with both hands, forming an enclosure around it. Opening and closing the

Figure 10-3 Micing the trombone
(Courtesy AKG Acoustics, Inc.).

Figure 10-4 Micing the clarinet
(Courtesy AKG Acoustics, Inc.).

hands helps control the volume. The instrument is quite small, so mic positioning can be critical.

Micing the trumpet.   The *trumpet* (Figure 10-6) is a wind instrument ending in a bell. It produces its tones by vibration of the player's lips against a cup-shaped mouthpiece. The instrument has valves that enable the use of all scale tones within its range. The mic does not face the bell directly. Instead, the sound is projected

Figure 10-5 Micing the harmonica
(Courtesy AKG Acoustics, Inc.).

Figure 10-6 Micing the trumpet
(Courtesy AKG Acoustics, Inc.).

slightly above the 0° to 180° axis of the mic. The working distance is about the same as the trombone.

Micing the tuba.   To record the tuba (Figure 10-7), mount the mic on a stand with its head facing down into the bell. Since the sound volume is large, start with a working distance of about 12 in. Either a cardioid or an omni mic can be used, but since the tail of the mic is pointed upward (away from the audience), possible noise pickup is reduced. The tuba is a large instrument, so it usually remains in a fixed position.

Figure 10-7   Micing the tuba (Courtesy AKG Acoustics, Inc.).

Recording String Instruments

The string instrument family includes the violin, the viola, cello, double bass, harp, acoustic and electric guitars, the banjo, and the ukelele.

The double, or string, bass.   The plucked strings of a *double bass* provide the fundamental tones and rhythm for an instrumental group. In recording low-frequency instruments such as a double bass with close mic placement, there is a tendency for recordists to confuse proximity effect in a single-element cardioid microphone with extended bass response. Two-way microphones do not exhibit proximity effect and can be effectively placed a few inches from the *f* hole on the upper side of the bridge of the instrument. A small personal microphone, such as a lavalier, can be wrapped in a layer of foam and placed in the opening of the upper *f* hole. The foam serves as a shock mount to minimize mechanical noise transmission to the mic. Figure 10-8 shows the tuning and pitch range of this instrument. Figure 10-9 indicates mic positioning.

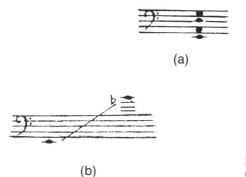

(a)

(b)

Figure 10-8   Tuning (a) and pitch range (b) of the double bass.

Figure 10-9   Micing the double bass (Courtesy AKG Acoustics, Inc.).

Micing the electric bass.   The *electric bass* (Figure 10-10), or fender bass, as it is sometimes called, needs amplification to join other instruments. A cardioid mic, or preferably, a two-way cardioid can be placed directly in front of the bass amplifier's speaker on axis with the cone. Mic attenuator pads will probably be required to prevent recorder or mixer overload. Another method is to use a transformer-type direct box at a point where the bass-guitar pickup feeds its amplifier for clearest sound. The transformer provides a stepped-down voltage suitable for direct connection to a mic input on the mixer.

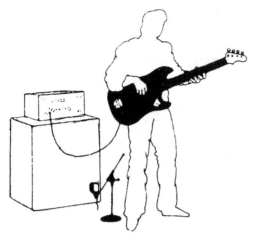

Figure 10-10   Micing the electric bass (Courtesy AKG Acoustics, Inc.).

Micing the acoustic guitar.   Guitars can be acoustic or amplified. The acoustic guitar has a larger body than that of the electric guitar and most of its sound is provided by its resonant body. A number of mic techniques can be used in recording this instrument. An omni mic placed about a foot from the circular opening of the guitar will give a good blend of guitar sound and the attack of the guitarist plucking the string. Cardioid mics must be used with caution, or proximity effect will provide an unnatural coloration of the guitar sound. A condenser mic provides clarity and detail to acoustical guitar performances. A special microphone contact device has been specifically designed for the acoustical classical guitar and may be useful in providing mic separation when recording a singing guitarist.

Although its resonant body is fairly long, as are its strings (compared to those of the violin), the sound output of the instrument is quite low, whether played with a pick or the fingers. As in the case of other instruments, there are a number of different guitar types, including the classical, the jumbo (or dreadnaught), and the *f*-hole arch top.

Figure 10-11 shows a method of mic positioning for this instrument. The mic is mounted on a stand and is pointed at the instrument hole or slightly below it, at a distance of several inches to 1 foot. There is maximum sound pickup with the mic in this position. Sometimes the musician will put a mic over the neck of the instrument but close to its body. The bass may sound more natural if an omni mic is used, but micing in close with a cardioid will emphasize low-frequency tones. In some instances the mic is mounted right on the edge of the instrument hole, held in position with a clip. The mic is an instrument type and is small so as not to interfere with playing.

To emphasize bass tones, bring the mic closer to the sound hole. The result is comparable to the proximity effect obtained by vocalists who mic in closely. Proximity effect is a characteristic of the mic. However, it is also caused by the

Figure 10-11 Micing the acoustic guitar (Courtesy AKG Acoustics, Inc.).

response of air volume inside the guitar with respect to bass tones. To get greater bass response, mic in closely and use either a cardioid or supercardioid. Results will also depend on the guitar. Sometimes this effect can be overdone, resulting in boomy sound.

Sometimes the acoustic guitar, because of its inherently soft sound, needs to be isolated to avoid mic spillover. Use a gobo and then move the mic closer to the bridge to emphasize treble tones or closer to the neck for a less bassy, more natural sound.

Mics can sometimes be used for sound mixing. In the case of the acoustic guitar use two mics, one near the bridge and the other close to the neck or between the neck and the sounding hole. This will result in two different sounds, and their proportions can be adjusted by moving the mics.

Contact mics can also be used to supply sound effects that are different than those supplied by boom-mounted mics. The result is sound that is not as natural.

Micing the electric guitar.   The output port of *electric guitar* has a high-impedance and so should not be directly connected to the low-input impedance of a tape deck. However, it can be done by using an impedance-matching transformer. The input of the transformer is plugged into the guitar; the output into the tape deck. The resulting sound could be called pure guitar. It is not modified and, assuming a quality tape deck, is undistorted.

A different technique is to connect the mic output to a fuzz box; then from the box to a wah-wah unit, foot-pedal controlled, and then connect the wah-wah to a guitar amplifier. The sound is then picked up by a mic facing the amplifier and positioned a short distance (about 1 foot, or possibly more) from it. The sound input to the mic can be strong enough to overload the preamplifier connected to the mic, producing enough distortion to destroy the guitar sound and possibly to damage the mic, especially if the mic is an electret or condenser type. A better mic for this purpose is the dynamic, since it is more rugged and can tolerate signal overload more readily.

Electric guitars can be recorded in a fashion similar to that for the electric bass, using a mic in front of the guitar amplifier's speaker or via a transformer-type direct box.

The guitar amplifier may contain two speakers. In that case do not position the mic an equal distance between the two. There may be some sound cancellation at this spot so move the mic off center. At the center position, the sound may be somewhat thin.

Recording two guitars and a solo vocalist.   As indicated in Figure 10-12, a single mic can be used to record two guitarists, with one or both musicians also acting as vocalist. The microphone is mounted on a swivel support so that the mic can be aimed higher or lower, as required. Although the mic isn't aimed at the instruments, this is somewhat overcome because there are two instruments.

Figure 10-12 Single-mic recording of two guitarists and a vocalist.

Whether to use one mic or two with either a single vocalist with one guitar or a pair of vocalists with two guitars depends on the effect to be achieved. The advantage of using two mics is that recording becomes easier, since each mic can be made independent of the other.

Mic location for solo vocalist with instrument.   It is common to record a vocalist using a guitar. But the voice and the instrument produce different sounds, and so different mics may be required. Further, as indicated in Figure 10-13, the instrument mic is placed so as to face the center hole of the instrument, while the mic for the voice is placed near the mouth, but slightly below it. The mics cannot be hand-held, so both are mounted on a vertical boom.

For recording, a cardioid is preferable to minimize off-axis sounds. The mic for the voice may be equipped with a three-position switch that allows dramatic

Figure 10-13 Two-mic pickup of vocalist with guitar (Courtesy AKG Acoustics, Inc.).

tailoring of the sound. These positions, identified as B, M, and S, are clearly marked on the mic body. B is the bass position and is used for low-frequency emphasis when the mic is worked close to the mouth; M is the medium position and supplies normal voice character without bass addition, with the middle and high frequencies somewhat accented for added vocal presence. S is the sharp position. The midrange and treble rise sharply to supply increased presence and brilliance. This is particularly effective when the accompaniment levels are very high.

Bowed String Instruments

Violins and violas have resonant bodies, which produce a combination of rich harmonic structure combined with the high-frequency harshness of the bowing action. A cardioid microphone placed to the violinist's side, aimed down from a boom slightly to the side of the bow, will give a rich sound with minimum harshness, depending on side placement.

Micing the violin.   When a violin string is bowed, several modes of vibration are set into action at the same time. The entire string vibrates, with the maximum amplitude at the center of the string at the fundamental frequency of the vibrating string. This frequency is determined not only by the length of the string but by its mass as well. The fundamental is inversely proportional to the length of the string; the longer the string, the lower the tone. The string will also have partial modes of vibration, with each half, quarter, eighth, and so on resulting in harmonics that are multiples of the fundamental frequency. This is applicable to harmonics lower than the seventh, but odd harmonics, starting with this one, are dissonant. However, as a general rule, these harmonics are progressively weaker.

Because the violinist may move the instrument toward and away from the microphone when playing, the microphone should have no proximity effect or artificial bass boost that could change the musical timbre of the instrument. Use a cardioid. Its pattern should be smooth so that the mic will respond without tonal coloration when the instrument is moved off-axis.

Try positioning the mic about two feet above the violin and direct it downward toward the bridge. With this location, there will be more freedom of movement of the instrument without change of violin timbre or volume (Figure 10-14).

Figure 10-14 Micing the violin (Courtesy AKG Acoustics, Inc.).

Micing the electric bass. The *electric bass* can be miced in the same way as the electric guitar, following the same precautions. If there are enough available tracks on the tape deck or mixer, try using both pickup methods, direct and mic, selecting the desired sound during mixdown. Alternatively, the two sounds can be mixed and then routed to the left or right channel or else to both channels to supply stereo bass.

Micing the harp. If the mic faces a harp player, then the *harp* sound can be very unbalanced. The low frequencies become booming, and the mid- and high frequencies move too much into the background. Consequently, it becomes convenient to position the mic behind the player's back and orient it over the right arm toward the center of the strings. This placing has to be done when the instrument is tilted toward the player. This mic location is more critical than in the case of other plucked instruments and should be checked by test recordings. Disturbing noises appear in the harp only during a change of tuning by pedals. It is also advisable to check this with a test recording as well.

Keyboard Instruments

The *piano*, the *grand piano*, and the *harpsichord* are characteristic of this group. With these instruments the strings are made to vibrate by a hammer or a plucking mechanism. Recordings of keyboard instruments can present problems. For the grand piano one reason for problems can be attributed to its great dynamic range. Recording results depend to a great extent on the room and on the positioning and direction of the mic. Although the sound of low frequencies propagates mainly omnidirectionally, the middle and high frequencies have distinctly directional characteristics, determined by the geometry of the instruments and the reflecting properties of the lid. As good rooms for piano recordings are rare, use of directional mics is almost always recommended. Place the mics so they are directed toward the descant strings. For solo recordings, use a greater working distance since a fuller sound can be achieved by including the reverberation of the room.

The piano is related to the drum in the sense that both are percussion instruments. For the piano, even with the help of a sustain pedal, a note begins to decay from the moment it is produced. The higher harmonics usually have a gradually decreasing amplitude with an increase in frequency and diminish first, and so the timbre of the instrument decreases smoothly with lower volume.

Micing the piano.   There are two types of instruments that are essential to music. One of these is the drum set, needed basically for establishing the beat. The second, and almost as desirable, is the piano, for it can not only supply the melody but can assist in the beat.

Not all pianos are alike. Aside from differences in structural quality, there are two basic types, and each of these requires a different recording technique. These differences are based on the position of the piano's sounding board: one is vertical, the other horizontal. *Upright pianos* and *spinets* use vertical sounding boards; *grand pianos*, including the baby grand and the concert grand, have horizontal sounding boards.

It would be unrealistic to expect uprights or spinets to supply the sound quality supplied by a grand. This has nothing to do with recording methods. If the basic sound source is inadequate, no amount of subsequent sound processing will change it. However, it is still possible to relieve the situation somewhat.

The music that is heard from a piano is supplied primarily by the sounding board. Although its dimensions are determined by the size of the piano, it can be augmented. One method for augmenting uprights and spinets is to move them a foot or two away from the wall. A typical wall will be moderately sound reflective and so will act as a supplementary sounding board. The sound reflected from the wall will strike other walls in the room, which will also act as a supplement to the piano's sounding board. This assumes these walls are not blocked and are reflective. Paneled walls are better than dry-wall types. Walls not blocked by chairs covered in soft material or a stuffed sofa are better. A solid wood floor, not covered by a rug, is helpful.

Moving the piano about a foot away from the wall results in a resonant air cavity, which helps reinforce piano sound, especially in the bass and treble regions. If the piano back is against a curtain or drape, try removing these decorations, which have treble-tone-absorption properties.

If the piano has a back panel, remove it. Also raise the lid of the piano. These two steps will increase the sound output. Further, it will be necessary to raise the lid to mic this instrument.

Unlike other musical instruments, which have a relatively concentrated sound source, the music produced by the piano is across a rather substantial length. For this reason it may be helpful to try a pair of mics, one for the bass and midrange section, the other for the midrange and treble section. Using a pair of mics also permits control of the relationship of bass to treble. This can be governed not only by correct positioning of the mics but also by the fader control in the recorder/mixer.

Mount the two mics on adjustable booms (Figure 10-15), with the mics pointing down into the interior of the instrument. Some experimentation is required regarding the actual positioning of the mics. Do not expect both mics to be precisely parallel to each other or for both to be pointing downward for the same distance.

The Old Upright Piano.   Although the keys may have yellowed, at least they are ivory and not plastic. Old upright pianos are larger than today's uprights and spinets and are capable of better sound. Their condition may require restringing, new key pads, or retuning. With some TLC, they can be sources of surprising quality. Mic them and prepare them for recording in the same way as spinets.

Figure 10-15   Micing the upright piano (Courtesy AKG Acoustics, Inc.).

The Tack Piano.   A *tack piano* is a modified upright or spinet that has been modified through the use of thumb tacks on the felt heads of the striking mechanism. The result is a sharp, metallic sound. The tacks need not remain imbedded permanently.

Micing the grand piano.   The grand and the baby grand have a number of points of superiority over the spinet and the small upright. They have a larger sounding board and the strings are longer, resulting in more resonant harmonics. Using a mic is easier, since it simply involves raising the lid and pointing the two mics down at the sounding board (Figure 10-16). Some experimentation with the adjustable lid may be helpful, with the underside of that lid working as a sound reflector. A contact mic, fastened to the sounding board, can produce some distinctive sounds.

Figure 10-16   Micing the grand piano (Courtesy AKG Acoustics, Inc.).

Grand pianos have a large string surface area producing sound. Microphone placement can be either monaural or stereo. For stereo recording, a cardioid mic placed near the upper third of the keyboard from the top will record the upper frequencies, and a second mic placed beneath the keyboard at the low register will record the lows. Placement and blending can provide a wide panorama of piano sound. Depending on room acoustics and microphones separation of instruments near the piano, a cardioid mic may be placed close for distinctive sharp piano sound or further away for better tonal blending. Generally, a single microphone placed in the middle of the piano's arc, aimed down at the strings, will give a good tonal blend. When recording a piano-playing vocalist, correct placement will assure separation of the vocal signal from that of the piano for best mixing and control.

Micing the harpsichord.   The *harpsichord* does not differ much from the grand piano as far as recording technique is concerned. Its dynamic range is, however, relatively limited. This must not, however, cause bringing the mic too close to the instrument, as the harpsichord produces substantial noise. If the harpsichord is played solo, obtain testwise balance between all the registers of the instrument and maximum suppression of interfering side noises.

In a chamber orchestra or in an instrument group, the harpsichord is not used solo. It is often unnecessary to allocate its own mic to this instrument. Recordings of an orchestra with a harpsichord present considerable difficulties. The ability to make a good recording depends to a great extent on the recordist's musical perceptions.

Percussion Instruments

Micing drums.   Drums include the bass drum, tenor and bell drum, and the tom-tom. Close placement of the mic is recommended to get a compact, precise sound character. Large drums with a bass drum pedal are preferably recorded from the skin side, opposite to the base pedal, to stop interfering mechanical noises. Basically, very close mic positioning on the clapper side is also possible. The snare drum is also recorded very closely. Be careful during play intervals of the small drum that the skin wirings, which cause a rattling sound, are damped or switched off. As a rule, cymbals of different types belong to the complete percussion group. These instruments are recorded relatively closely, obliquely from above.

A minimum of two, but for better results, three mics are needed for the optimum recording of percussion instruments.

Micing the kick drum.   There are two factors that will influence the recorded sound of the kick drum. Even when positioned in an identical manner, a pair of mics may produce different results. If you have a number of mics available, try them all to determine which is most suitable.

Remove the drum head and position the mic inside the drum. Or, place the mic on top of a blanket inside the drumhead for the characteristic thump-thump-thump sound preferred by many professional drummers. If there are rattling or buzzing problems, put masking tape across the drumhead to damp these. An alternative approach is to use a pad of paper towels. To keep the towels from unraveling, hold them in place with Scotch tape. A cardioid should work well with this instrument.

Micing the drum set.   The *drum set* is a number of instruments, as indicated in Figure 10-17, consisting of a snare drum, tom-toms, splash cymbals, a hi-hat, and a bass drum. The drum produces a sound whose level is enough to overwhelm one or all of the other instruments.

How the drum set is miced depends on what you expect from this combi-

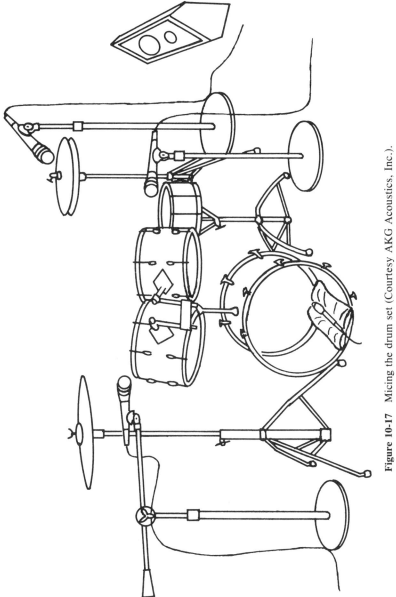

Figure 10-17 Micing the drum set (Courtesy AKG Acoustics, Inc.).

nation. Unless all the music to be played follows somewhat the same pattern, it may be necessary to move various mics around or to fade some of them. If the preference is for getting a sharp drum sound, use a condenser mic having an omni pattern, mount it on a boom and have it pointing down at the drum. This arrangement supplies a sharp attack if used in combination with a peak limiter.

There are a number of ways of micing the drum set, depending on how much ambient sound you want to include. Try a cardioid for minimum sound, or an omni for maximum response. If background noise, either side-to-side or 180° off-axis, is a problem, stay with the cardioid.

Another mic position to try involves positioning the mic at the side of the drum head, close to the drum's skin surface. With this arrangement, it becomes possible to pick up a larger number of harmonics.

The drum set is probably the most difficult part of a musical combo to record, possibly because it isn't a single instrument. Further, these drums cover a much wider frequency range than the average single instrument, ranging from a low of about 50 Hz to a treble of 3 kHz or more.

Drums can be troublesome for a number of reasons. Depending on the drum, the sound frequency can be deep bass, which is difficult to record. Drums are percussive and so are characterized by strong transients, short-lived waveforms having very high amplitudes. Since these last a very short time the much slower ballistic-type VU meters on the mixer may not even indicate them. When drums are used, try recording below 0 VU.

One technique is to record each instrument of the drum set individually. Equalize and use the fader at the mixing console. In addition to the separate mics, mount one on a boom to record the overall drum-set sound and to pick up room reverb.

If a number of instruments are to be recorded ensemble, use cardioids to minimize leakage. If leakage isn't a problem, try omni mics so as to cover as many of the instruments as possible.

The bass drum can be miced in several ways, producing different kinds of sound. The mic can be positioned facing the skin or, more often, putting the mic inside. Two techniques are used. The mic can be positioned off-center to avoid standing waves. Another is to pad the interior with a blanket, a method that helps discourage standing-wave formation.

Drum-set sound has a large number of variables, and so the best way to get a good recording is to experiment. The final drum sound depends on the expertise of the performer, the age and type of instruments, the number and types of mics, and their positioning. There is no formula to follow that will guarantee specific drum-set sound.

A cardioid is an excellent choice for difficult staccato-type drum sounds, and if the mic has a built-in bass roll-off, this will allow a bright sound for more drum presence. If a tight drum sound is wanted from the kick drum, use a mic that is a shade less responsive to bass tones.

Micing the drummer-vocalist.   A drummer-vocalist may use as many as four microphones. For voice reproduction, the microphone should have its bass response somewhat attenuated to avoid possible excessive pickup of lower frequencies from the drum set. An advantage would be reduced proximity effect and a lower possibility of feedback problems.

A cardioid is recommended for the vocals. If the mic has excellent front-to-back discrimination, it will be fairly insensitive to unwanted sound from the rear and sides. Low-frequency roll-off helps reject the low-frequency noises found in less-than-ideal performance situations, such as air-conditioning noise, noise from electric fans, footsteps from an upper floor, or conversation from the rear. It also helps keep out unwanted drum sounds.

Kettle drums or very large drums in the orchestra are not usually provided with individual mics. On the contrary, it is advisable that any large and booming instruments should not be too strongly recorded by nearby mics.

Micing triangles, gongs, xylophone or vibraphone, and celesta. These instruments do not cause particular difficulties in recording. However, in rooms with a longer reverberation time, short, sharp strokes and too great a mic distance can cause an impression of rhythmic irregularity. It is generally recommended that mics be positioned relatively closely to these instruments, but at the same time to ask the players to reduce the range of sound volume.

Micing congas.   Use an identical pair of mics to record *congas*, with each boom-mounted. Position the mic heads so they point downward toward the center of the skins with a starting working distance of about 8 in. (Figure 10-18).

Figure 10-18 Micing congas (Courtesy AKG Acoustics, Inc.).

Micing bongos. Use a single boom-mounted mic pointed downward toward the center space between the two instruments (Figure 10-19).

Figure 10-19 Micing bongos (Courtesy AKG Acoustics, Inc.).

ELECTRONIC INSTRUMENTS

Electronic musical instruments are usually reproduced by loudspeakers. This does not mean that each of these instruments has its own built-in loudspeaker. Many electronic organs, synthesizers, electro-guitars, or even the sound tape recorder itself must be connected with loudspeakers through external power amplifiers to make the electronically generated sound audible. Under certain circumstances such instruments can be recorded directly—that is, without the use of a mic. When the instrument delivers just the signal voltage, it is advantageous to transmit this voltage by cable directly to a tape recorder. When micing an amplifier system followed by a loudspeaker, remember that the woofer, midrange, and tweeter are often housed in one enclosure and, further, that several of each speaker type may be included. It will be helpful to remove the grille and to become aware of the number and type of speakers and their direction of firing. They do not always fire in a forward direction.

Micing pipe and electronic organs. Possibly more than with any other instrument, room acoustics play an important part in the sound produced by these instruments. The pipe organ does produce instrument-operation sound, which is noticeable when the mic is close, including both air and mechanical sounds.

It is desirable to use several mics for organ recording. The mics can be cardioids positioned close to the instrument's pipes or, for electronic organs, close to the instrument's speakers. Some recordists use three mics, with two stereo mics working as the main pickups, one at the left and the other at the right. The third mic is used to pick up reverberant sound. For a pipe organ, using close left and right mics helps pick up tonal harmonics.

Recording Instrumental Groups

For recording the free sound field of an instrumental group, such as a quartet, use a pair of mics separated by 1 foot. The two mics are suspended from a horizontal crossbar, with the crossbar held in place by a single-element boom, which can be adjusted to raise or lower the mics. Although both mics can be stand-mounted on a stereo bar, they can also be hung from a cable yoke on a stereo bar from 12 feet to about 15 feet above the floor level. They should be pointed down at the musicians at an angle of about 45°.

When micing string groups, some professional musicians prefer a mic with a slightly attenuated bass response for a clear string sound. Experience has indicated that a tailored bass response in a mic and a slight rise in both the upper-midrange and higher frequencies project an instrumental presence that adds tonal clarity while diminishing the possibility of environmental muddiness.

An electret condenser microphone is widely used by professional musicians. An electret cardioid adds virtually no coloration to sounds that reach it off-axis. Not all electrets are alike. A top-performing unit will have a long, trouble-free life.

VOCAL AND ORCHESTRAL RECORDING

The human voice is one of the most expressive musical instruments. Even when the acoustics of a recording studio are less than ideal, the proper use of a microphone can compensate for them.

Positioning a microphone correctly is an art, not a science. For every instrument there are some generalized instructions, but often enough these are starting guidelines. The results depend on the characteristics of the microphone, studio acoustics, and some willingness to experiment. The advantage of experimentation is that it leads to experience.

Angle of Incidence of the Microphone

Professionals never sing directly into a microphone because they know it will pick up excessive breath noise (aspiration), which usually happens when the mic is 1 inch from the mouth of the performer. It requires just an additional inch or two of working distance to eliminate aspiration, but this can cause sibilance. It is better, as indicated in Figure 10-20, to sing to one side of the microphone or over its head.

The hand-held cardioid, rather than a stand- or boom-mounted mic, is best for solo vocalists. The mic can be held relatively close to the mouth, thus supplying

Figure 10-20 A vocalist should sing to one side of mic or above it.

a higher S/N ratio. This, plus the cardioid's polar pattern, provide less opportunity for the pickup of ambient noise. It is also helpful for the elimination of popping, supplied by consonants *p* and *t*. The hand-held mic gives the soloist an opportunity to get better separation and control of any accompanying instruments. Vocal soloists using a hand-held mic have an advantage over musical instruments, which must use fixed-position microphones. An experienced performer can adjust working distance to accommodate soft or loud musical passages.

Micing the Small Orchestra

The faithful reproduction of the sound of a small orchestra calls for planning. For stereo reproduction, some professional audio technicians are sometimes inclined to use only two microphones to reproduce full orchestral sound. Two identical microphones positioned on a crossbar will give full coverage to various instrumentalists while maintaining proper orchestral perspective.

In this recording situation, a cardioid mic is a fine choice if it is free of the proximity effect that is usually found in such mics. The mic should have a uniform cardioid pattern, one that supplies virtually no coloration of off-axis sound sources. This is an important consideration when the mic must reproduce the sound of various instruments situated in relatively different positions.

Micing Big Orchestral Sound

One way of learning how to reproduce the sound of a large orchestra is to learn it from professionals, those who developed their expertise from actual recordings, not only in studios but in a number of halls, chambers, and auditoriums as well as outdoors.

In a typical symphony orchestra arrangement, where the sound of a large number of musicians must be captured, one technique—referred to as *flying the mic*—not only can be effective but also is the easiest approach. Physically, the mics are out of the way and are almost unnoticeable visually.

THE RECORDING STUDIO SETUP

There are a number of possible recording studio arrangements. Although they are all electronic, there are a number of variations under this heading. The studio can be completely analog, from the microphone input to the monitor output. It can be a combination of analog and digital, including a synthesizer in its recording chain. Or, it can be almost completely digital, omitting the microphones and using one or more synthesizers controlled by a MIDI. Even here, with an emphasis on digital components, the output must ultimately be brought into headphones and/ or monitors, and those components, as well as our ears, are designed for analog reproduction. The difference, then, is in the amount of digital equipment used. The great advantage of a digitally oriented studio is its ability not just to play various effects, but to put them into a computer memory and to call them up when needed.

With digital electronics, it is possible to have synthesized single-instrument music, but, by using a number of synthesizers and a synthesized drum combo, it is also possible to have a studio that is almost completely digital.

Digital does not supplant analog; it supplements it. Although analog methods are useful for recording of known sounds, digital permits the creation of entirely new sounds and new sound combinations, and so the sound engineer becomes not only a sound recordist but a sound innovator as well.

The recording studio is as much a piece of equipment as any microphone, monitor, mixer, or any other electromechanical device. In a sense, it is more difficult to achieve a top studio than to have good equipment, because good equipment can be bought; it is provided with specs, and what it can (or cannot) do is no great mystery. This is not the case with the selection and setup of a studio. The best studio, from a sound viewpoint, is one that is initially as sonically inert as possible. It should be nonreverberant and as noise-free as possible.

These are the two basic problems—uncontrolled reverberation and uncontrolled noise. Reverberation can be controlled by using materials whose specs indicate a high percentage of sound-absorption capability. These materials do not prevent the transmission of sound from the outside to the inside of the studio. For that, a material such as brick or concrete is suitable. Both materials have different jobs and both are needed.

MULTITRACK RECORDING

As its name implies, *multitrack recording* involves recording different sections or individual instruments of an orchestra on tape, but recording them separately. A complete orchestra need not be assembled, and sections or individuals can be recorded at separate times. This does not imply that such recording is done on a haphazard basis.

In such recording a reference is required, which is usually a signal consisting of the first musical beat. Recordists differ in their approach to multitrack recording, but the drum set may be used at the start to supply the beat. The rhythm and bass are used as the foundation for dubbing in the melody. How the tracking is done also depends on the number of tracks that are available for recording.

Multitrack Recording with Memory

A computer memory can be used as an aid in multitrack recording. It is first used for storing a variety of voices, instruments, and rhythms, any of which can be recalled as required.

There can be many such storages, but these are not done on magnetic tape; rather, they are put in a computer's electronic memory. It is also possible to experiment with sound combinations, adding vocals and/or instrumentation to the initial tracks and storing these separately. The various tracks can be mixed down, ultimately recorded on tape. They can also be monitored throughout the entire process. Further, the original memory tracks can be overdubbed.

The Recording Concept

The recordist has two options. In concert-hall recording, the acoustics of the hall are regarded as part of the music. Sometimes substantial efforts are made to tune the hall by modifying it acoustically, but the fact remains that the sound can still be different, depending on a person's seating position. The resulting sound is governed not only by the architectural acoustics but also by the number of microphones, their location, and their distance from and their angle with respect to the performing instruments. The number of persons in an audience and the kind of clothing they wear also helps determine the acoustics of the enclosure. As a result, acoustic control is much more difficult in an auditorium than it is in a recording studio.

Recording in the home is somewhere between recording in an auditorium and in a studio. However, in-home recording usually does not have the benefit of artificial reverb, so a simulation of a concert hall is not to be expected.

Acoustics can be controlled more easily in a studio. The effort here is often to make the recording room as passive as possible and then to introduce artificial reverb to give an impression of a large or small volume of space. The noise level is generally reduced in the studio, since an effort is usually made to do so.

STUDIO REQUIREMENTS

Effect of Room Acoustics

The acoustics of any room—including in-home rooms, concert halls, auditoriums, small or large recording studios, and the confined listening spaces of automobiles, trucks, or boats, are an integral part of the recording and playback process. There is one possible exception, that of sound playback outdoors, but even here there is the possibility of overcoming the reverberation-free environment by using headphones.

Room acoustics are just as much a part of the recording process as any microphone, mixer, or any other equipment. The word *coloring* is sometimes applied to the effect that acoustics have on a recording. Poor acoustics result in sound that is muddied, both during recording and playback. Acoustics are not an enemy, but they must be controlled just as any other recording instrument must be controlled.

The Desirability of Noise

It may seem strange to regard noise as desirable, but—like it or not— we live in a world of noise and are so accustomed to it that a completely noise-free environment would seem strange and disturbing. Noise need not come from some external source, since the human ear supplies its own. Usually its level is governed automatically, and we aren't conscious of it. An uncontrolled condition is known as tinnitus, in which the level can be intermittent, ranging from a mild hiss to a roar. There is no cure for persons suffering from this condition.

Extraneous noise in a studio is undesirable, but it can be controlled to a lesser or greater degree. In a worst-case situation it is possible for noise to mask music.

Sound Coloration

The geometry of a recording studio, its size, the angles of the walls with respect to each other, the absorption properties of materials in that room, the number of windows (if there are any) and whether they are open or closed, and the number of mics and their positioning all have an effect on sound frequencies. Because of these working conditions, some of which are fixed, others variable, the reverberation in a room may prefer certain sound frequencies. The effect is that some sounds are strengthened and are reinforced, leading to the effect known as *coloration*, an effect that does not appear in the original musical script. Coloration is not only caused by the reinforcement of certain sound frequencies but also by standing waves. For a recording studio, both frequency-selected reverberations and standing waves must be eliminated.

Standing Waves

Also known as stationary waves, these are resonances that do not move and are due to the particular dimensions of the studio but can also exist in a playback situation in the home. Sound that moves through the air forms progressive waves, but these can result in standing waves when they are incident on a reflecting surface and are reflected along the same patch as the incident wave.

Studio Echo

An echo is the delayed reflection of sound. There are various types, often affected by the size of a studio. For small studios, there may be a number of echoes, having a somewhat small amplitude and coming in rapid succession. These are sometimes referred to as *flutter echoes*. In a large studio an echo may be a single, but distinct, type. In either instance, echoes are undesirable and interfere with recordings.

However, there are instances in which echoes are desirable for producing certain sound effects, particularly in broadcasting. A room made specifically for the production of the hollow sound effect of an echo is known as an *echo chamber*. Such rooms are also used for imitating the natural reverberation of a room, such as a concert hall or auditorium, with this simulated sound used to give the recorded sound a feeling of these enclosed environments.

An echo is determined by its reverberation time, which is at least 0.05 second but can range to as high as 2 seconds. In some instances an echo can be artificially induced by using a device known as a reverberation plate.

Sound Diffusion

The objective in a sound studio is to achieve a diffusion of both direct and reverberant sound. Ideally, the sound level should be the same in all parts of the room. No architectural feature should have the ability to focus the sound. This means that no surface—walls or ceiling—should have convex or concave shapes. The function of a studio is to consider sound to be of prime importance, with architectural attractiveness of no consequence. Good sound diffusion is hindered by standing waves and echoes.

There are a number of advantages to a studio characterized by good diffusion. It makes good microphone placement much easier. It permits easier placement of the performers. And, most important, it makes it more possible to achieve desirable recordings.

Absorption Coefficient

The absorption coefficient is a measure of the sound energy absorbed when dry, or direct, sound strikes a reflecting surface. The absorption coefficient depends on the angle of sound incidence and on the frequency of the sound. The frequency includes not only the fundamental but all harmonics as well.

Acoustical Tiles

Some materials, such as acoustical tile, are made specifically for this purpose and have a high coefficient of absorption. The absorption coefficient of acoustical tile ranges from 0.8 to 0.9, indicating that most of the sound is absorbed, with very little reflected. Highly polished marble, on the other hand, has a sound absorption coefficient of 0.01, indicating that 99.99% of all incident sound is reflected.

Studio walls and the ceiling as well can be covered with acoustical tile. This is a porous material and, at frequencies above 1 kHz, is very absorptive. The tile is made of pressed particle board or fiberglass.

Acoustical materials are also measured in terms of their density. This is the ratio of weight to volume. In the English (customary) system, the measurement is in pounds per cubic foot; and in the CGS system is in grams per cubic centimeter.

Sound Diffraction

Depending on architectural construction, sound can be bent. Bending can be caused by a small space, the edge of a wall, or by a corner. Diffraction modifies the sound when it passes the edges of an opaque body or strikes an uneven surface.

Mass Law

The ability of a material to reduce the transmission of sound is proportional to its weight. According to this law, to reduce the transmission of sound by 6 dB, the thickness of the wall must be doubled.

STC

STC is an abbreviation for *sound transmission class* or *coefficient*. It is used for making a comparison of the sound transmission characteristics of different materials used in the construction of a sound studio.

Transmission is a reference to the propogation of sound through any substance that has sound incident upon it. The extremes of sound transmission range from 0% to 100%. A completely open space, such as an open window, has a *sound-transmission coefficient* (STC) of 100%, excluding the window frames or any sound-reflective materials associated with the window.

Sound-Level Meter

A *sound-level meter* is a sound-pressure-sensitive component used for the measurement of loudness. It is equipped with several scales, with each corresponding to the sensitivity of the human ear at different sound frequencies.

RECORDING STUDIO COMPONENTS

The first components in a recording studio are the microphones; their characteristics have been described in Chapter 4. Their selection and positioning have also been discussed earlier in this chapter. Following the microphones the next component is a mixer.

The Mixer

The input to the mixer can be directly from the microphone(s), with this input identified as mic/line input. An alternative input would be from a tape deck, and this can be either a cassette, an open-reel unit or a compact disc player. In any case, the input signal to the mixer is electrical and is selected by a switch.

The Mixer Input

Following these two possible inputs, the signal is brought into an audio amplifier, referred to as a *preamplifier* (preamp). The preamp is an audio amplifier, and its function is to bring the input signal up to a desired level.

Sound input can be controlled in two ways. One method is to move the microphones so as to obtain a suitable sound level and mix from the instruments. Using the mixing console is much easier for it can combine the outputs of the various mics and at the same time adjust the different levels from each. Still another technique, one that is used in multitrack recording, is to record the output of each mic on a channel of multitrack tape. These channels can be handled individually, with their levels raised or lowered as desired.

RECORDER/MIXERS

Aside from their specs, recorder/mixers are characterized by the number and versatility of their features. These units can be divided into three categories: The first is the in-home recorder/mixer, which is the lowest-cost unit, has the least number of features, and may be marginal as far as quality is concerned. It is limited insofar as inputs are concerned; it usually has just two and is a two-track type. The second type is the median recorder/mixer. This unit occupies a niche between the professional type and the in-home component. It has more features than the in-home but fewer than the professional. Its quality may be satisfactory but not suitable for those insistent on top performance. It will probably have four inputs and a two- or four-track recording capability. It may be equipped with features such as an effects input, may or may not have reverb, and will have either tone controls or some form of equalization. Costwise, it is somewhere between the basic in-home recorder/mixer and the professional. The professional is the top-of-the-line unit, may have as many as eight inputs, is equalizer equipped, and is designed for special effects. The mixer and recorder can be separate units, as in Figure 10-21.

There are, of course, variations within each class, and although recorder/mixers have been put into three categories, there is no sharp demarcation line between the top-of-the-line of one class and the bottom-of-the-line of the following class.

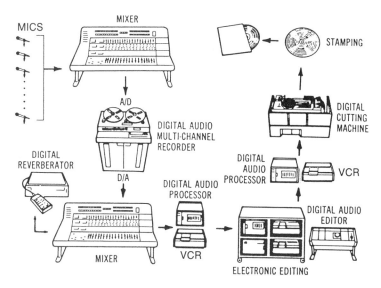

Figure 10-21 Mic output is fed into mixer and then into an analog/digital (A/D) multichannel recorder (Courtesy Sony Corporation of America).

Functions of the Recorder/Mixer

The name of this component, whether used at home or in a studio, supplies a guide to its function. It records and it mixes. In its recording function, it puts the sounds of one or more musical instruments or vocalists onto one track or onto more than one. As a practical expedient, the beat is recorded first, including the drum set and rhythm instruments such as the bass.

Using headphones, the recordist can then overdub, a technique in which new sounds are put on tracks that have not yet been used. The overdubbed material consists of accompaniment instruments, possibly a violin, flute, cornet, vocal, or others.

Mixdown is the final step. This is the stage following the recording of all the instruments. No matter how many tracks have been used in prior recording, they are all combined into two tracks, one for left-channel sound, the other for right-channel sound. The mixdown tape can be a cassette or open-reel tape. This is regarded as the master, for it is from this tape that prerecorded cassette tapes (usually) or open-reel tapes can be made.

For professional studios the recorder/mixer is not one, but a number of components housed in a single enclosure. These components can consist of an input module, actually a group of controls used for making adjustments to a single signal. The module itself, as shown in Figure 10-22, is just one of a number of similar modules. Usually a recorder/mixer has a limited number of modules, from as few as two to as many as eight. The greater the number of modules, the greater the number of mics that can be used.

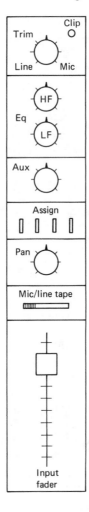

Figure 10-22 Mixer input is controlled by an input module.

Mixer Input Impedance

Some mixers have low-impedance inputs; others have inputs that are rated high. All that matters when inputting microphones is compatability for a mic to produce its best frequency response and sensitivity. Even if initially the impedances do not match, it is simply a matter of using the appropriate transformer and connecting cable.

Functioning of the Module

The *module* can be considered the control center of the recorder/mixer, but its function will vary from one component to another. It tends to have more functions in recorder/mixers that have more operating options.

Clip

The word *clip* refers to clipping, or cutting, of the peak amplitude of the audio signal and is a form of distortion. Clipping is evidence of an overly strong audio signal. In the recorder/mixer it consists of a light-emitting diode (LED), which lights or flashes to call attention to a clipping condition. If allowed to continue, the excessively high signal will overload the input of the following preamp. The input fader can be used to reduce signal strength to the point where the clip LED ceases flashing.

Trim

The *trim* control is a variable resistor, or pot (potentiometer), whose function is similar to that of a volume control in a radio receiver. Like the input fader, the trim control is used to adjust the level of the input signal. It is a coarse control when compared to the input fader.

If the input signal has a wide dynamic range, it may be necessary to ride either the trim control or the input fader. Use the trim control for obtaining a coarse adjustment, to the point where the clip LED does not light or flash. To obtain optimum input signal strength, ride the input fader to the setting at which the clip LED starts to flash, and then back off. The VU (volume unit) meters can also be used for evidence of possible signal overload, but ballistically these are not as sensitive as the clip LED. In some instances the clip LED will flash before the VU pointer reaches its overload point.

Mic Control

The controls shown at the top of the input module include settings marked mic and line. If the input signal is to be supplied by one or more mics set the control to the mic position. The line signal is one that could be supplied by a tape deck, an electronic synthesizer, or an electric guitar.

These two possible inputs, line and mic, can have different signal-voltage levels. The signal supplied by a mic will have a range of about 0.001 to 0.002 V (1 to 2 mV). The line signal is higher and has a range of approximately 0.3 to 1.23 V.

Equalization

In one type of recorder/mixer, the module has two equalizer controls, one of which is marked LF (low frequencies) and the other HF (high frequencies).

There are two types of equalizers, graphic and parametric, and both belong to the tone-control family. The graphic equalizer is a less expensive unit and typically ranges from a three-band to a five-band unit. The three-band type divides the audio spectrum into three bands: low, midrange, and high frequencies. The five-band divides the audio spectrum into five bands and so is able to supply finer control. Some of the more elaborate graphic equalizers divide the audio spectrum into as many as 25 bands. The equalizer control allows a change in volume of the audio band it covers by about ±12 dB.

The advantage of the graphic equalizer over a tone control is that it covers a much narrower range and so is able to increase signal strength or decrease it over this range without affecting adjacent frequencies, above and below the selected range.

The parametric equalizer permits control of the width of the selected portion of the audio spectrum, so that it can be narrow or wide. The parametric equalizer may include a variable-frequency control and variable boost or cut. Like the graphic equalizer, it may include both left- and right-channel coverage.

Aux

The *aux (auxiliary)* control in the module in Figure 10-22 is shown directly below the equalizer controls. It is adjusted to determine the amount of signal level that is routed to an external effects (FX) device, such as a time-delay or reverb unit. Just a portion of the signal is sent to the external effects component, and after it is processed, it is returned to the main signal.

If no processing is needed, the aux control is kept in its zero position. Processing may be needed if the original signal, mic or line, is too "dry."

The module used here shows just a single aux control (sometimes labeled as aux send), but a more elaborate recorder/mixer may have two. Some of the less sophisticated components are not equipped with an aux feature.

The aux controls have an additional function, which permits checking on the progress of the signal and also to supply a signal mix. In this case the aux isn't used to sent the signal to an effects component, but to a small amplifier connected to a pair of headphones.

Signal Routing in the Recorder/Mixer

Depending on how it is generated, whether it is mic, line, or tape, the signal is brought into a preamp, then through an input fader, and then into an equalizer. From the equalizer the signal is routed to mixing circuits via a pan pot (potentiometer). At this point the signal is assigned to odd and even tape tracks. The signal is then routed to amplifier mixing circuits. Each amplifier has a master feeder that is VU meter equipped. The mixdown supplies a pair of tracks, one of which is identified as track 1, the other as track 2.

Faders

The recorder/mixer will have several faders, either slide-control or knob-equipped. One of these will be an input fader, the other a pair of master faders. The input fader is positioned between the signal preamp and the equalizer. The master faders follow the pair of mixing circuits. Consequently, there are faders preceding and following signal mixing.

There may also be a direct output connection immediately following the input fader. The direct output connection leads to a jack, so that the quality of the signal can be checked prior to mixing.

Faders are members of the attenuator family, and as such they behave as losser networks. The fader in Figure 10-23(a) is a continuously variable device; and in this example is used to control the output voltage of a pair of mics.

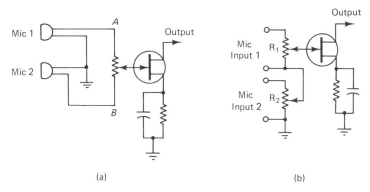

Figure 10-23   Fader controls. (a) Fader does not provide mixer action. (b) Mixer-type fader.

The fader is grounded at its electrical center, and when the sweep arm of the potentiometer is in this position, the voltage input to the following transistor amplifier stage is zero. When the arm is at point *A*, maximum input from mic 1 is supplied to the amplifier. Similarly, when the arm is at point *B*, there is maximum input from mic 2. Input to the amplifier is supplied by either mic 1 or mic 2, but there is no simultaneous input from both mics. As a result, this circuit does not supply sound mixing action.

Mixer-Fader

The fader circuit can be modified to supply mixing action, as shown in Figure 10-23(b). This circuit has a pair of independent variable resistors, with the voltages developed across them wired in series-aiding; hence they are additive. A microphone is connected across input 1 and another across input 2, with the amounts of these voltages determined by the settings of R_1 and R_2.

Volume Unit Meters

Volume unit meters (VU meters) are used as volume indicators on recorder/mixers. VU meters are voltmeters and are calibrated in decibels with respect to a reference voltage. VU meters are made specifically for monitoring complex sound voltages and in a recording studio are used as indicators on recorder/mixers. To emphasize 0 VU, the scale is deliberately made thick from that point to the extreme right, or 3 VU. 0 VU indicates a reference level of 1 mW across 600 Ω; with this as a reference, the meter is a dBm type.

VU meters frequently have two scales. The upper scale is in black from the left side of the meter to the 0-VU point and red from that point to the extreme right. The lower scale can indicate percentage of modulation. Full-scale deflection accuracy is 2%.

LINE LEVEL

Line level is the signal-input amplitude to a recorder/mixer and is the level that produces 0 on a VU meter. It is also known as the 0 level, the reference level, or the nominal operating level, in addition to line level. 0 on a VU meter is specified in volts dBv or dBm. dBv is a term used with mics when output level is referenced to 1 V. dBm indicates a level with reference to 1 mW in a 600-Ω line.

Not all recorders use the same nominal operating level. A quarter-track multi-track unit would have a nominal operating level of − 10 dBv (0.316 V). For a large multitrack half-track format, the nominal operating level is specified as 0 dBm (0.775 V).

However, no matter what specified operating level is used, these are still higher than the output of a mic, which is usually 0.001 V for a dynamic type and 0.003 V for a condenser mic. The reason for the higher voltage output for the condenser mic is that it uses a built-in impedance-matching transformer, which also supplies a voltage step-up. For this mic, as well as for the others, a voltage preamp is used to bring the output level up to a suitable point for acceptance by a recorder.

Line Input

As its name implies, *line input* is a connecting point for the output of a mic preamp or other high-level devices such as equalizers or compressor/limiters. An attenuator pad, working in conjunction with an overload indicator, is used to control the signal level on the recorder.

RECORDING AND DYNAMIC RANGE

The sound level of a symphony orchestra can reach 105 dB SPL or momentarily higher with loud signal peaks. A rock group can produce even-higher amounts of SPL, sometimes approaching sound levels that result in ear pain. In this sense music is capable of interfering with itself, since the higher harmonics of musical tones generally (but not always) have lower amplitudes than the fundamentals with which they are associated. In short, a harmonic can have a higher amplitude than its fundamental.

To be able to record all the sound of live music, the recording medium should be able to accommodate a dynamic range of at least 100 dB. There are two facts that make even a higher dynamic range desirable. One of these is the peak achieved by some initial transients present at the beginning of a waveform. The other is the use of percussive instruments such as the piano, drum beats, and cymbals resulting in momentary, very high signal peaks.

Both cassettes and open-reel tapes are used in recorder/mixers, but even a reel-to-reel type at a speed of 7-1/2 ips has a capability of only 60 dB dynamic range. This is the measured S/N ratio of a high signal level that results in 3% harmonic distortion. The result, though, is audible signal degradation, which indicates that while 60 dB may be supplied in the specs, it is not a clean 60 dB. A safe procedure is to subtract about 10 dB so as to supply an adequate safety margin below the point of audible distortion and above the noise floor. Thus the tape recorder is required to record musical programs having a dynamic range in dB nearly twice its own capability.

The result of this disparity is that either the top 50 dB of the music will be distorted, with the recording driven into tape saturation, or the bottom 50 dB of the music will be buried in the noise floor of the tape. An alternative possibility is that some of the music will share the noise floor of the tape at the bottom end of the dynamic range of the tape recorder and also the distortion region at the top end of the dynamic range.

Dynamic Range Reduction

One solution is the intentional reduction of the music's dynamic range during recording; this can be done in several ways. The orchestra's conductor can ask the musicians deliberately to restrict the loudness of their instruments and to increase the sound level of the softer passages. A more common method is for the recording engineer to ride the gain control, thus manually controlling the dynamic range of the signal being fed into the recorder. Another technique is to use electronic gain or level controls called *compressors* and *limiters*. A compressor reduces the dynamic range by gently reducing the gain of loud signals and by increasing the gain in the presence of quieter signals. A limiter acts more drastically to restrict any loud signal that exceeds some preset level.

Musical compositions with a lower dynamic range are considerably easier to record correctly. It is necessary only to make certain that at the strongest fortissimo of the composition, any overmodulation is avoided. This can be done by a prior test for the peak-level setting of this fortissimo before the beginning of the recording itself. As a rule, no follow-up adjustment is then necessary, provided that even at the lowest passages of the composition concerned, the recording-level indicator is still just deflecting.

SOUNDPROOFING

Soundproofing an in-home room or modest-budget recording studio is essential for several reasons. Good soundproofing reduces the presence of noise and to determine the amount of noise it is essential to listen for it consciously. A truly soundproof room is noise-free. It is also nonreverberant and supplies dry sound only.

Artificial reverberant sound can be added; its advantage is that it can be controlled and adjusted until its effect on the dry sound is specifically what is wanted by the sound engineer.

Soundproofing Techniques

Any room intended for studio use, whether in the home, semiprofessional, or wholly professional, can be soundproofed to some extent. Even limited sound-proofing is better than none at all. Some soundproofing methods are simple and require no expenditures.

1. Close all room openings. Keep windows closed and make sure they are airtight. If the room has a window fan, remove or cover it thoroughly. If the room is cooled via vents, close them or cover them during recording time. If possible, use a room for recording that does not require the presence of people, fans, or air-conditioning. The sound-transmission coefficient of air is 1, which is the highest of all substances.

2. Receptacles (outlets) are openings into walls; they possibly back up receptacles on other walls. They do permit the passage of air. Use padding behind the wall plates of such receptacles. Some pads are made to fit directly behind the wall plate.

3. Make sure all doors used for entry into the recording studio are solid. Doors are often made with 1/8-in. sheets of wood mounted over a door frame. Most of the door is air and so permits the relatively easy passage of sound. Replace such doors with solid types. Both open-frame and solid doors attenuate sound, but the solid type can supply an additional 10 dB.

4. Do not build a studio in a room containing a fireplace unless you are willing to cover the opening with carefully fitted solid material. The same is true of a room having a pass-through or an archway.

5. Weatherstrip all doors and windows. Do this not only around the outer perimeter of the windows but also across the center portion containing the window lock. As a test, light a candle on a windy day and check to make sure no air comes through. The flame should not flicker.

6. Use storm windows. Make sure these are airtight and, as an added precaution, line them with window-sealing materials. Add materials such as felt or fiberglass to the open space. This has a double purpose. It helps insulate the room and it damps any possible window resonances.

7. Put carpeting and underpadding on the floor of the room above the recording studio. This will help minimize the sound of footsteps and other unwanted sounds and will help damp resonances. This carpeting will have a double function, because it will not only keep above-studio-room sounds out, it will minimize sound from the studio reaching the upper room. Warm air in a room tends to move upward, carrying sound with it.

8. If the ceiling has not been acoustically treated, cover it with sound-absorbent tiles. Use the heaviest tiles obtainable, but even lightweight tiles are better than none at all. Acoustic tiles aren't noted for their mass, and so most of the sound energy from a room above will pass right through to the studio. A better approach is to replace tiles with boards of some kind, such as plywood or fibreboard.

9. Keep bric-a-brac out of the studio. Flower vases, glass ornaments, wall- or shelf-mounted pictures, and floor and table lamps have a natural resonant frequency. All these can interfere with a recording. Sometimes these resonances are difficult to trace.

10. Keeping windows closed and sealed and heating vents shut tight can make the studio extremely uncomfortable in summer and winter. If you need air-conditioning, use a conditioner mounted on a table rather than in a window. Enclose the unit in a box that is padded, but use a rear vent and front vent to allow the movement of cooled air.

The Need for Sound Reflection

Sealing a studio against the intrusion of outside noise can have an unfortunate side effect—reducing reverberation. This can be overcome through the use of an artificial reverb unit. A studio can be shielded against noise, but increasing the area of reflective surfaces can permit reverb to reach a satisfactory level.

ARTIFICIAL REVERBERATION

As far as reverb is concerned, there are two possibilities when recording in a studio. One is to use the natural reverb; the other is to take advantage of artificial reverb. Natural reverb has a distinct disadvantage, since the studio may have been acoustically modified to minimize or eliminate noise. The problem is that the attenuation of noise can interfere with reverb, thus reducing the liveliness of recorded sound.

If the studio is acoustically dead, the dry sound can be recorded and then the desired amount of reverb can be added. Although reverb can be supplied acoustically, electromechanically, or purely electronically, the electronic method is easier and better in the sense that it can supply a greater variety.

For home recording, the technique used is to make the recording room as acoustically dead as possible and then to add artificial reverb.

Artificially induced reverberation, whether it is barely noticeable or in depth, is essential for recording studios, broadcast stations, theaters, and individual performers. In every case, reverb supplies control over an enclosed environment and helps make the recording engineer master of the sound environment. Artificial reverb is used in the range from mix auditioning to final mastering. It can enhance, or "sweeten," individual instruments, vocalists, or an entire group. Even in live performances, it can create variations in depth and fullness of sound through a spectrum of controlled spaciousness.

In broadcast studios, a reverb can be used with live voice to increase the average modulation level, to increase station "loudness," and to increase the signal-coverage area. It can also provide enhancement and special effects in the production of commercials.

There are various features that are desirable for a reverb unit. It should have a high density of resonant frequencies; a high pulse density to duplicate the many sound paths of naturally reverberant environments; a high degree of diffusion in both frequency and time domains; a linear frequency response for a maximum range of applications; a precise duplication of natural-room reverberation effects (no flutters, pings, etc.); an adjustable input sensitivity; built-in limiters to prevent overdriving of inputs; built-in reverb/dry signal mixing; no positive feedback, even when placed close to monitor loudspeakers; and equipped with standard three-pin audio connectors for all inputs and outputs.

There are a number of different kinds of reverb, and those available depend on the versatility of the multieffects processor. It is possible to get reverb suited to a hall, a smaller room, or a room suited for vocals or the type of reverb produced by a plate reverberator or an echo room. An echo room is one in which the studio engineer has extensive control over the room's dimensions and other room parameters. It is possible to have a reverberation program that includes a programmable gate that can be selectively triggered by the input signal, a separate analog signal, or by a trigger foot switch.

Early Reflections

The concept of time does not enter into dry sound. If there are no sound reflections, the dry sound is heard uniformly. However, with reverberant sound, the first few reflections that occur just after a sound is produced and before the reflections become dense enough to be called reverberation are known as *early reflections*. These early reflections add presence to vocals and instruments. As indicated in Figure 10-24, the reverb signal decreases in amplitude, dropping 60 dB from its start to finish.

Gate reverb combines early reflections with a programmable, triggerable gate, which shapes the sound effects as required. The reverse gate produces a contemporary reverse-reverb effect that can sound like a tape being played backwards.

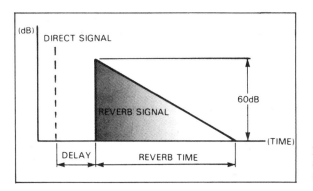

Figure 10-24   Reverb amplitude decreases with time (Courtesy Yamaha Corporation).

THE DIGITAL MULTIEFFECT PROCESSOR

In the studio a digital multieffect processor can work with a MIDI keyboard, with an electric guitar, and as part of a dual-sound reinforcement system. It can offer as many as 30 preset programs, ranging from reverb through a reverse-gate effect to parametric equalization.

The term *preset*, as used here, could possibly supply a wrong impression. Each program can have up to nine parameters to help control the precise sound that is wanted. It can also use as many as 60 RAM memories for storing personally created and edited programs.

As an example, the digital multieffect freeze programs permit sampling up to 2 full seconds of continuous sound. The maximum initial delay time for reverb effects has a full 1 second capability. The modulation capabilities of effects such as stereo flange, chorus, and stereo phasing are enhanced for greater control. The sound-processing features described here include some of those offered by a fully equipped digital processor.

Delay

The *delay* effect offers independently variable left- and right-channel delays, adding spatial interest to a fairly common effect and permitting the creation of a dubbed sound. The maximum delay time for this effect is a full 2 seconds.

Echo

Using the multieffect device, *echo* has independently programmable left- and right-channel delay times. The application of feedback creates a gradually diminishing series of repeats that constitutes the familiar echo effect. It is possible to obtain independent programming of left- and right-channel feedback levels, making it possible to create extremely complex echo effects.

Modulation

Modulation effects are produced by periodically varying the amplitude, frequency, or delay time of an input signal. Using the digital multieffect device makes it possible to obtain effects such as stereo flange, chorus, stereo phasing, symphony, and tremolo. It can also dramatically thicken the sound of any instrument.

Chorus.   The *chorus* feature serves to thicken the sound of an instrument, creating the effect of several instruments playing at the same time.

Symphonic effect.   *Symphonic effect* is a multichorus effect and is therefore quite dense.

Phasing.   *Phasing* is basically a gentler, swirling effect, lending a smooth, animated quality to the original sound.

Tremolo.   *Tremolo* effect supplies a fluctuation in the volume of the signal, ranging from subtle to extreme.

Pitch Change

Basically, a *pitch-change program* alters the pitch of an input signal in semitone increments over a two-octave range, plus or minus one octave. Fine adjustment in 1-cent increments is also possible (1 cent = 1/100 semitone). Pitch change permits detuning an instrument against itself to create subtle chorus-like effects, thus imitating complete ensembles with a single instrument. The pitch-change interval can even be remotely controlled from a MIDI keyboard or sequencer.

The amount of pitch change available depends on the processor. Some are capable of shifting the input over a four-octave range (plus or minus two octaves). Pitch change makes it possible to produce two independently pitch-shifted notes in addition to the direct signal, making it possible to create three-part harmonies with a single input note. Both pitch-shifted notes appear at the center of the stereo sound field.

Freeze

The *freeze* feature is essentially various sampling programs that permit recording of up to a 2-second sound segment in the memory of the multieffect device. The entire sound or a specific portion of it can then be played back as required. A variation of the freeze program even permits changing the pitch of the sampled sound from a MIDI keyboard, so it is possible to play the sample as if it were a keyboard voice.

Sound Compression

Compression limits the dynamic range of an input signal. It does this by boosting soft signals and reducing high-level signals to keep them within a specific range. This is an effective way to smooth the dynamics of a bass guitar, to increase the sustain of a guitar, or to tame a jumpy vocal performance. This is the effect produced as a vocalist moves closer to or away from a mic. It is also used to limit the maximum level of a signal in order to prevent input overload.

The expander function spreads the dynamic range of a signal to allow efficient suppression of low-level noise, resulting in a cleaner-sounding overall signal.

Harmonic Addition

Harmonics add to the fullness of a tone and help make instruments more readily identifiable. The digital multieffects unit can artificially add appropriate harmonics to the input signal. This helps bring buried sounds to the foreground without making them louder.

Noise Gate

This feature uses a gate to pass or shut off the input signal in a number of ways. It can be used to pass just a short segment of a longer input signal, or it can be set up to pass only signals that exceed a specified level. In the latter case, it functions as a *noise gate*, gating out the noise when no signal is present. It can also be used to create reverse-gate-type effects in which the signal level increases gradually after the effect is triggered.

TIME-DELAY EQUIPMENT

With time-delay equipment, there has been a gradual replacement of analog by digital techniques. Primitive acoustical tube delays, with their inherent group-delay distortions and low-frequency cutoffs, were initially replaced by so-called bucket-brigade delay lines.

While the recorder/mixer is the heart of a sound studio, there are other components, sometimes referred to as *add-ons*. These can include sound-processing devices such as noise reduction units, time-delay devices, digital reverberation, and so on.

Delays for Reverberation

Many devices that produce artificial reverberation start to reverberate too early and produce no initial reflections, which are essential for the perception of room characteristics. To overcome this disadvantage, the onset of reverb is delayed and the delay gap is filled up with one, two, or three reflections per channel.

Accent Microphone Delays

Another application for a digital delay add-on is the compensation of the propagation difference between accent mics and the stereo mics in live performances. Sometimes, special attention is brought to a soloist by delaying the performer's mic a few milliseconds less than the rest.

Pseudo Stereo

Pseudo stereo is a technique that broadens the image of two speakers driven by a mono source. This is achieved by splitting the mono signal into two parts by means of a positive and a negative flanger of fixed delay. The complementary frequency responses of the flangers distribute the spectrum of the mono signal evenly between the two speakers.

Digital Delay

A digital time-delay unit is useful for sound reinforcement in a special application, such as a long room that uses as many as eight loudspeakers. A distributed loudspeaker system with progressive amounts of time delay improves the intelligibility and gives the illusion that the sound comes from the original source.

The block diagram in Figure 10-25 shows the connections of the digital time-delay unit to the mixer. The input signal is stereo, with the output connected to

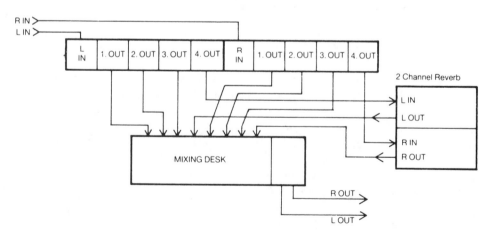

Figure 10-25 Connections for add-on digital time-delay unit (Courtesy AKG Acoustics, Inc.).

the mixer via the wiring from a panel on the rear apron. The L and R signals from the mixer are supplied to the stereo reverb section of the add-on unit. The added reverb is then routed to the mixer.

The technique used in digital delay starts with a sampling of the audio signal. The sample is converted to a series of binary digits (bits) in a manner similar to that of quantization. These bits are then stored in digital shift registers or memories. The time between the original storage and the time when the bits are called out of storage is the desired amount of delay. Following the exit of the binary digits, a series of 0s and 1s, they are converted to audio by using a D/A converter.

This technique has a number of advantages over tape or analog delay units. There is no incurred distortion during storage time. The sound quality can also be controlled by the number of bits—that is, the quantity of binary numbers. Depending on the time allotted to storage, the effect can be echo (time-delayed reverbertion) or a true reverberation effect.

The time difference between echo and reverberation is measured in milliseconds. For an echo, that time is about 50 milliseconds; for reverb it is less than that.

SOUND EFFECTS

Various sound effects can be produced by techniques known as flanging, wah-wah, fuzz, and sounds can be produced to simulate background noises, such as thunder, the opening or closing of a door, the rattling of a window blind, horses hooves, and so on. Musicians can also produce a variety of sounds through instrument technique.

Flanging

Flanging is done mechanically by lightly pressing on the supply reel of an open-reel tape deck. This slows the rotation, thus modifying the reel speed and lowering the sound frequency. Alternatively, the take-up reel can be fingered to make it move a little faster, an action that increases the frequency. A better method is to use an electronic flanger. This controls the tape speed, above and below normal, in a manner that can be repeated more precisely than with the fingers. This is advantageous if the same sound effect is to be repeated at more than one place in the recording.

Stereo flange.   The *stereo flange* effect results in a swirling aircraft sound that is popular with guitarists. The flanging effect is produced by varying the delay between two identical signals, thus producing a complex, varying comb-filter effect.

Wah-Wah

An electric guitarist can produce an effect known as *wah-wah* through the use of a control known as a *wah-wah pedal* connected to a special-effects box, which, in turn, is connected to one of the input terminals of the guitar amplifier. The wah-wah sound can be introduced at any time during a composition.

Fuzz

Fuzz, designed to accompany the playing of an electric guitar, is produced in a manner similar to that of wah-wah. The guitarist has a foot pedal for controlling a special-effects box. The signal output of the box is fed to one of the inputs to the guitar amplifier.

Distortions such as wah-wah and fuzz have long been used with electric guitars, as add-on components for these instruments. A multieffects device also offers these sounds. The distortion program is capable of producing an extremely broad range of guitar-style distortion sounds. In addition to distortion levels, it offers many equalization parameters that supply a wide range of overdrive effects.

Panning

Panning is the movement of the sound image between left and right in the stereo sound field. For electronic multieffects, pan direction, speed, and depth can be programmed for instant recall. Some multieffects devices allow the creation of rotary pan in addition to standard stereo pan effects. The sound seems to sweep toward and away from the listener, in addition to having left-to-right or right-to-left movement. A triggered pan program automatically pans the sound image across the stereo sound field, with programmable attack, pan, and release rates.

ADR Noise Gate

The *ADR (automated dialog replacement)* uses a gate to pass or shut off the input signal in a number of useful ways. It can be used to pass just a short segment of a longer input signal, or it can be set up to pass only signals that exceed a specified level. In the latter case, this program can function as a noise gate, excluding noise when no signal is present. It is also possible to create reverse-gate-type effects, in which the signal level increases gradually after the effect is triggered. Some multieffect devices supply a foot switch, and it is possible to use this aid to trigger the gate.

Two-Channel Programs

It does not follow that an effects program selected for one sound channel must automatically be used in the other. A multieffects signal processor can provide different, independent effects for the left and right channels. The output may be mixed and delivered in stereo, or each channel can function independently as a mono processor. The programs can include plate plus hall reverb, early reflection plus reverb, echo plus reverb, chorus plus reverb, and two different types of panning.

Multiple Effects

In addition to two-channel programs in which separate effects are used for the left- and right-sound channels, an effects processor can also offer a combination of effects used by the two channels simultaneously. This program is a combination of compressor, distortion, equalizer, dynamic filter, reverb, chorus, and symphonic or exciter effects, with a range of programmable parameters for each effect.

MULTIPLE-MICROPHONE INTERFERENCE

The advantage of using more than one mic, in addition to being able to supply stereo sound, is that several mics can create the equivalent of a more uniform sound field. But unless the mics are correctly positioned with respect to each other, it is possible to develop sound frequency cancellations.

A basic desirable setup is shown in Figure 10-26(a). Here the mics are positioned 1 foot from the sound source and are 3 feet from each other; this positioning is sometimes referred to as a 3-to-1 ratio. However, this fundamental positioning of multiple mics is often ignored, which results in some strange sounds during sound reinforcement or recording.

Figure 10-26(b) supplies some examples of poor microphone positioning. The first illustration shows mic pickup from four different sound sources. Although vocalists are shown in the drawing, the sound could be that supplied by instrumentalists as well. Here the mics are spaced 1 foot apart from each other and are also placed 1 foot from the source. Another poor arrangement is that shown in the next drawing, in which a pair of mics are separated by $1\frac{1}{2}$ feet and are angled inward at about 45°, with each mic at a distance of 1 foot from the sound.

Figure 10-26(c) shows the use of two mics for recording four sound sources. Here the 3-to-1 rule is observed, since the mics are separated by 3 feet and have a distance of 1 foot from the sound sources. The mics aren't angled, but the sources are.

The arrangement in Figure 10-26(d) can be used when a pair of mics are selected to pick up a double source. The mics are angled outward and are positioned 1 foot from the sources. In all these illustrations it is assumed that identical mics are used and if they are equipped with frequency controls these all occupy identical settings.

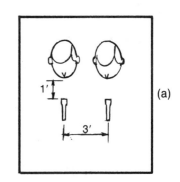

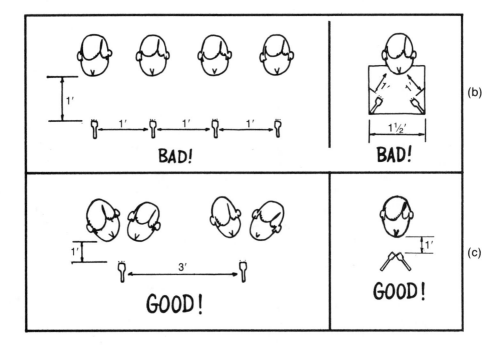

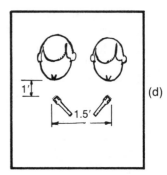

Figure 10-26 Microphone interference and techniques for reducing it (Courtesy Electro-Voice, Inc.).

The Gobo

Recording a group of musicians can be done in several ways. The entire group can share one mic, and in that case both left- and right-channel sounds are recorded simultaneously. An alternative technique is to record individual instruments or small groups and assign their sounds to selected tracks on the recorder/mixer. When the mixer contains enough tracks, instruments can be recorded individually.

When instruments are recorded in groups and a number of microphones are used, there is the problem of sound spillage, with the sound from one instrument being picked up by the mic for some other instrument. This is a type of mixing, but is undesirable because it takes sound control away from the sound engineer.

There are two ways of minimizing sound spillover. One is to use a cardioid mic with the maximum sound on axis with the mic. Close micing is used. Another method is to use a gobo, a solidly built frame (or group of frames) for enclosing the player and the instrument. The gobo encloses the instrument on three sides, permitting reverberant sound to enter from the fourth, open side.

The gobo must be substantially built and enclose as much of the instrument as possible. To keep its dimensions to a reasonable figure, the performer should be seated. The gobo can be constructed from particle board, $\frac{3}{4}$-in. plywood, or masonite. Sandwiching these materials is desirable, although it contributes substantially to the weight. Blankets, cotton, and old clothing forced between wooden panels is not as satisfactory as a sandwich of solid wood. Make sure the gobo rests on the floor and since the gobo consists of separate sections, their edges should butt against each other. An amazing amount of sound can leak through even a very small opening. The trouble with a gobo is that it will tend to block some reverberations and will make the recorded instrument sound rather dry.

In some recording situations, it may be necessary to use: cardioids, micing in closely, and working with gobos as well.

Reverberation can contribute to sound leakage from one mic to another. Another way of minimizing this is to use heavy curtains over all windows. When recording, extend the curtains as widely as possible. These will absorb some sound, reducing reverberation and giving the sound a dead effect. This can be overcome in subsequent sound processing. If artificial reverb isn't available, it will be necessary to experiment with the gobos and curtains until maximum reverb is available with minimum sound spillage from one mic to the next. Still another technique is to separate the performers and their mics as much as possible.

If there is a choice of several rooms to use as a recording studio, select the largest. This will allow maximum separation of mics and will also produce a smaller number of standing waves.

Index